SOTHIS PRESS

Sothis Press aims to completely reimagine the public domain book with an emphasis on universal access to captivatingly designed print editions. We believe that technology will enable every single book ever published to be perpetually accessible in beautiful print form. Sothis is beauty in ubiquity.

www.sothis.press
1 Waterhouse Square, London, ECIN 2ST

The First Crossing of Spitsbergen

Being an Account of an Inland Journey of Exploration
and Survey, With Description of Several
Mountain Ascents, of Boat Expeditions
in Ice Fjord, of a Voyage to North-East-Land,
the Seven Islands, Down Hinloopen Strait,
Nearly to Wiches Land, and Into Most

William Martin Conway

SOTHIS PRESS

ISBN 978-1-0365-0151-8

The First Crossing of Spitsbergen
is in the public domain.

Being an Account of an Inland Journey of Exploration and Survey, with Descriptions of several Mountain Ascents, of Boat Expeditions in Ice Fjord, of a Voyage to North-East-Land, the Seven Islands, down Hinloopen Strait, nearly to Wiches Land, and into most of the Fjords of Spitsbergen, and of an almost complete circumnavigation of the main Island. By

SIR WILLIAM MARTIN CONWAY

M.A., F.S.A., F.R.G.S., Sometime Roscoe Professor of Art, University College, Liverpool

With Contributions by J. W. GREGORY, D.Sc.
A. TREVOR-BATTYE, and
E. J. GARWOOD

Together with Eight Coloured Plates reproduced in facsimile from Sketches by H. E. CONWAY, Two Maps, and about One Hundred Full-Page and Text Illustrations from Photographs and Sketches

MDCCCXCVII

LONDON — J. M. DENT & CO.
67 ST. JAMES'S STREET, S.W., AND
ALDINE HOUSE, E.C.

Printed by Ballantyne, Hanson & Co.
At the Ballantyne Press

PREFACE

THE maps which accompany this volume were engraved for the Royal Geographical Society, and are here reproduced by its permission. Two illustrations made from drawings by Mr. A. D. MacCormick, my old Himalayan companion, originally appeared in the "Alpine Journal," and I am indebted to the Editor of that publication for permission to use them. I must also not omit to thank Mr. Trevor-Battye for the drawing of Wiches Land which is here reproduced.

W. M. C.

CONTENTS

CONTENTS

LIST OF COLOURED PLATES

MAPS

LIST OF ILLUSTRATIONS

The Illustrations marked with an asterisk are reproduced from Drawings by Mr. H. E. Conway.

CAMP SCENE.

SPITSBERGEN

CHAPTER I

INTRODUCTORY

IT was in Lord Dufferin's "Letters from High Latitudes" that Spitsbergen[1] first emerged, for me, from the fogs and darkness of Arctic mystery, as a land of mountains and glaciers, of splintered peaks and icy bays, a place worth seeing and even worth going to see. It is, as every one knows, the portion of Arctic land which has been more frequently visited than any other, for the simple reason that of all Arctic lands it is the most accessible. The same Gulf Stream that renders our own islands so temperate, so wholesome, and so damp,

[1] Not *Spitzbergen*. The name is Dutch.

pushing its warm waters towards the Pole, melts in the ice-covered sea a bay of open water. This bay extends in summer to the 80th and sometimes even to the 82nd parallel of north latitude, and thus forms an exceptionally easy avenue of approach towards the polar regions. Spitsbergen skirts, through several degrees of latitude, the eastern side of this open bay. The waters of the Gulf Stream impinge, it is said, upon the long mountainous island, named (in King James I.'s days) Prince Charles' Foreland, then pass round its northern and southern ends and open ways to the actual coast of the main island.

The name Spitsbergen properly applies only to this main island, along whose western margin stand a series of mountains composed of hard archæan rocks, often splintered into sheer and striking peaks, whereof the reader of this volume, it is hoped, will derive some idea. Associated with the main island are a number of others forming an archipelago. Three are of considerable size. North-East Land, whose position is indicated by its name, is the most remote. Edges Land (otherwise called Stans Foreland) and Barendsz Land are separated from one another and from Spitsbergen by very narrow channels, and indeed practically form its south-east limb. The remaining islands are small and numerous—the summits of submerged mountains or table-lands resembling similar portions of the neighbouring land.

Some fifty-five miles east of Barendsz Land and twenty-five miles south of North-East Land is a group of rather large islands properly called Wiches Land, but now generally known as King Carl's Land. These we had the rare good fortune to approach very closely, a thing seldom possible. Somewhere to the east of North-East Land is likewise an island or group of islands named Gillis Land, not known to have been attained by man. Our attempt to gain sight of

Gillis Land was no more successful than other attempts repeatedly made by our best equipped predecessors.

The history of the exploration of this interesting archipelago is a topic abounding in novelty, and so large and important that I hope soon to devote a separate volume to its sole consideration. The materials collected are already so numerous, and the subject presents such extraordinary ramifications and developments, that time is needed to pursue the study with needful thoroughness. At present I merely note in this place a few well-known facts, suited to throw light on the following narrative, that the reader may understand the purpose and plan of our journey.

Spitsbergen was discovered by the Dutchmen Barendszoon and Heemskerk on the 17th of June 1596. They were at the time sailing northwards to find a way over the Pole from Holland to China. In 1607 the same coast was revisited and further explored by the English navigator Hudson, sailing with a purpose similar to that of Barendsz; but Hudson observed the prevalence of whales, walruses, and other valuable animals, and fisheries were immediately established by Englishmen in consequence. During the first quarter of the seventeenth century the Spitsbergen waters became the scene of much international rivalry, the English attempting to annex the land and secure a monopoly of the fisheries, whilst foreign "interlopers" of various nationalities successfully resisted their pretensions. Ultimately a working arrangement was made between the parties concerned; the harbours and bays on the west and north coasts were divided between the rival fishermen, the Dutch taking Fair Haven and Dutch Bay (which was the best whale-fishing base), the English Magdalena Bay, English Bay, and so forth.

The prosperity of the bay-fishery did not last long, for the whales presently abandoned the bays and had to be sought in the open sea; but while it lasted Spitsbergen was a very busy and populous place during the summer months. Buildings were set up for habitation and blubber-boiling. Crowds of people assembled at the various centres. As many as 18,000 are stated to have made Smeerenburg at one time the centre of their operations. By the middle of the seventeenth century the whale industry was already declining, and a few years later Smeerenburg was a vanishing ruin.

Whalers, however, continued to visit Spitsbergen with diminishing frequency till about the year 1830 or even later. When they no longer came to seek whales or boil blubber, they landed for water, or to secure supplies of fresh meat from the quantity of easily obtained reindeer found in all the fertile localities. The walrus and seal industries for some time outlasted that of whaling, but now walruses have become practically extinct in the accessible parts of Spitsbergen. The only walruses we saw were on the edge of the ice-pack in Olga Strait.

About the middle of the eighteenth century many Russian trappers from the Arkangel district made the archipelago the scene of their activity. They used to spend the winter there, building for themselves scattered huts, the ruins of which are still discoverable. At first they also did very well, but in time they exhausted the supply of bears and foxes which formed the staple of their catch. About 1830 the last of the Russians disappeared to return no more.

They in their turn were succeeded by Norwegians, who now alone make these islands and waters the home of any industry. Sloops and cutters from Hammerfest and Tromsö still visit Spitsbergen in small and perhaps decreasing number, and there endeavour to secure a mixed cargo of whatever they can take, eider-down, seals, white whales, sharks' livers, a bear or two, perhaps a few walruses from North-East Land, but chiefly reindeer, the meat of which is sold in Norway at a good price. Thus Dutchmen, Englishmen, Germans, Biscayans, Russians, and Norwegians have all at one time or another sought Spitsbergen for industrial purposes, and by their ruthless methods of extermination reduced it to its present almost lifeless condition. Unfortunately it continues to be a no-man's land, annexed by no state and governed by no laws. Fisheries are unregulated; there is no close time for bird or beast, and so the animal depopulation threatens to become complete. In the interests of science and industry alike it is time Spitsbergen were annexed by some power capable of regulating the country. The Norwegians are the people upon whom the task should fall.

I have often been asked what the inhabitants of Spitsbergen are like. There are no inhabitants, and never have been any, if the few Russian trappers are excepted who spent some consecutive years in the island. Its shores have proved inhospitable to attempting colonists. Samoyedes could doubtless thrive there, but no one has ever tried to introduce them.

The scientific exploration of Spitsbergen has been the work of the present century. I do not refer to the employment of the island as a base for polar exploration by Parry and others, but to the investigation of its form, its geology, its fauna and flora, its climate, and its glaciers. In 1827 Keilhau, a Norwegian, began the study of Spitsbergen geology, but Professor Sven Lovén, who visited it ten years later, is to be regarded as the real originator of its systematic scientific exploration. He was followed by Otto Torell and Nordenskjöld in 1858, between which year and 1896 Sweden sent no less than nine scientific expeditions to Spitsbergen. With these the name of Nordenskjöld is most prominently associated. Meanwhile neither England nor Germany was idle, as the several voyages of Lamont, Leigh Smith, and Von Heuglin sufficiently attest.

These expeditions, with hardly an exception, confined their attentions to the coasts and outlying islands. A ship was the most convenient base, and few were the occasions when explorers ventured more than half a day's march inland. I find it recorded by Lamont[1] that "some years before 1869" a party of wrecked walrus-hunters travelled on foot overland from the Norways to Cross Bay and wintered in Möller's Bay, but nothing is known of their route or what they saw. In 1873, Nordenskjöld, after wintering in Mossel Bay, worked round the north coast of North-East Land; landing on the east coast, he crossed the great sheet of ice covering the whole interior, and reached Hinloopen Strait by way of Wahlenberg's Bay.

In 1890, Gustaf Nordenskjöld and two companions landed in Horn Sound and made a rapid traverse over the inland ice on snow-shoes to Recherche Bay, whilst later in the season they went overland from Advent Bay to Coles Bay. Lastly, in 1892, Monsieur Charles Rabot, having only forty-eight hours at his disposal, landed in Sassen Bay and made a plucky attempt to find a way across to the east coast by following the Sassendal. He reached the mouth of the fourth south side-valley (our Turn-back Valley), and climbed the hill beyond, to which he gave the name Pic Milne-Edwards.

It is thus evident that, up to the year 1896, the interior of Spitsbergen was practically unknown. The island had never been crossed, whilst such descriptions of its nature as had been given, by persons who looked inland from high points of view near the coast, were, as might have been feared and as we afterward proved, altogether misleading. When I began to study the literature of Spitsbergen topography, nothing surprised me more than the manifest indifference of travellers to everything concerning

[1] "Yachting in the Arctic Seas," p. 238.

the interior, an indifference perhaps characteristic of yachtsmen and seagoing folk in general. A German visitor, for instance, who climbed to a high point on Mount Lindstrom, near Coles Bay, described the view inland to the south as being over an "unabsehbare weisse Flache." Other writers spoke of the hills in the same neighbourhood as being the fronts of a great plateau. Only Mons. Rabot and Herr G. Nordenskjöld gave truthful and intelligible accounts of the kind of country they saw. Various writers spoke of having landed and advanced up valleys in pursuit of reindeer; but it seems never to have occurred to any one of them to note the bearing of the valley's direction, still less the position and number of side-valleys. When they added estimates of the distance they advanced inland, to which it is possible to apply tests, the estimated distances always turn out to be ludicrously exaggerated.

Thus it came to pass that, after taking the best advice we could obtain, we equipped ourselves with Nansen sledges, and ponies to draw them. It was believed that we should have to drag our things for a few miles over soft bogs, and that then we could find smooth areas of snow over which advance would be rapid and easy. It was the central portion of the island that we were to explore, the northern and southern portions being supposed to be wholly buried under great ice-sheets, though, as we afterwards proved, there is much mountainous country and many green valleys in the neighbourhood of Wijde Bay. In the central portion of the island many valleys were recorded as penetrating the hills. By one or other of these we imagined it would be easy to gain access to some snow-covered plateau continuous to the east coast. Even Mons. Rabot thought we should have to cross such a plateau east of Peak Milne-Edwards.

We had not been a week on the island of Spitsbergen before we discovered the utter unsuitability of Nansen sledges for the work we had to do. We ought to have brought Samoyede sledges. With them we could have accomplished easily what we accomplished only as the result of the most toilsome exertions, and perhaps we might have done more in our time. Such, however, are always the drawbacks under which pioneers labour. Learning from and profiting by our experience, a party may go in some future year and add largely to our knowledge of this most interesting island. The ponies, again, were a great anxiety to us. We decided on taking them because of Mr. Jackson's favourable account of their usefulness in Franz Josef's Land. But he had Icelandic ponies. We were obliged to put up with the larger and less hardy Norwegian beasts. The first that were supplied to us were unsuitable and had to be sold at a sacrifice in Trondhjem. At Tromsö we acquired better animals, which served us well, but they gave much trouble, and the question of how to feed them was always a difficulty. In the concluding chapter of this book I shall record the result of our experience, as far as it is likely to be useful to future explorers. Suffice it here to say that Nansen sledges, while excellent for ice-work, are the worst for boggy and stony places; whilst ponies, which are most useful in bogs and valleys, are practically valueless on crevassed, snowy, and icy areas. Our combination of ponies with Nansen sledges was therefore about the worst possible.

The reader must bear in mind that the main object of our journey was to cross Spitsbergen and reveal the character of its interior. Before seeing the island we thought the method to pursue would be to strike across the island along two or three lines. After three days spent in the country we found that a different method must be chosen. The intricate nature of its topography involved detailed study of a specimen area, and this we accordingly undertook and carried through.

The science and topography of the coast found no place whatever in our plans. Zoologists, geologists, photographers, and topographers—we were all pledged to one another to subordinate every consideration to that of getting inland. Our plans found approval at the hands of the Council of the Royal Geographical Society, which voted us a handsome subsidy, whilst the Royal Society placed a sum of money at my disposal from the Government grant to enable me to invite Dr. J. W. Gregory of the British Museum to join our party, and the Trustees of that great national institution gave him special leave of absence. How valuable was the companionship of the author of the "Great Rift Valley of Africa," how useful was his experience, how helpful his energy, readers of that fascinating description of a most plucky and fruitful journey will readily appreciate. Mr. A. Trevor-Battye, author of "Ice-bound on Kolguev," was likewise good enough to share our fortunes. He came as zoologist, intending to pay special attention to the fauna of the interior. Unluckily there was little for him to observe, for the birds chiefly frequent the coasts. Whilst we were busy inland he was able, by using the walrus-boat, to visit the little-known North Fjord and Dickson's Bay, where his work was compensation for the otherwise serious loss of time entailed upon us by lack of a boat.

Mr. E. J. Garwood likewise joined the party, and thereby contributed a most important addition to our strength. He came as geologist, photographer, and expert mountaineer, and in all these functions did admirable work. As a climber he had won a first-rate reputation amongst the élite of the present generation, having accomplished numerous expeditions of the first order of difficulty, and being accustomed to dispense with professional assistance. He proved himself a quick and accurate geological observer, and an excellent traveller, full of energy, helpfulness, and kindness. To him belongs the credit for the two most important mountain-climbs of the journey—those of Mount Starashchin and Horn Sunds Tind—which were either accomplished by him alone or under his leadership. My cousin, Mr. H. E. Conway, came as the artist of the expedition. He worked industriously and with no little skill in most uncomfortable surroundings and ungenial weather. The present volume contains many of his notes, whilst his sketches in water-colour and oils have already been exhibited to the public. Lastly, we were joined for a few days by Mr. J. T. Studley, who came in the capacity of a sportsman. Finding nothing to attract him in the appearance of the interior, he left us, to our no little regret.

Our first plan had been to secure a steamer for a shifting base, and thus to attack the interior from various points in succession. Had we been able to carry this out we should have profited, but the expense of a steamer proved too heavy for our means, and no generous yachtsman was forthcoming. The boat must have been large enough to take the two ponies on board, so that our choice was restricted. We thus had to find some other means for reaching the island. In answer to my advertisement a Norwegian firm communicated with me, stating that the enterprising Vesteraalen Steamship Company intended to build a visitors' hut in Spitsbergen that very summer, and that it would be set up at Advent Bay, the point I had already selected for our base. Accordingly I agreed to share with them the steamer that was to take up the building and the workmen who were to erect it, they undertaking to carry our ponies, our open walrus-boat, and our two Norwegian followers, and to fetch them back to Norway at the close of the season. The agent whom I employed to make this arrangement unfortunately bungled the matter and committed us to return three weeks earlier than we had intended. The tourist-hut itself was of course useless to us.

It went up with us in pieces, and was still in that disjunctive condition when we began our journey into the interior. We were then absent from the coast for thirty-six consecutive days. During this time, whenever we did not sleep in the open air, our only shelter was a tiny tent not tall enough to stand up in, which barely kept out the rain and not at all the cold. On our return I only remained in camp at Advent Point long enough to overhaul the baggage and obtain possession of the 12-ton iron steamer in which the five of us, cooped up with all our belongings in a cabin without floor-space for all to stand up at one time, voyaged over a thousand miles, often in the midst of heavy ice, and along coasts, through straits, and up bays for the most part never before visited except by properly-built Arctic vessels, and not often even by them. Returning to Advent Point, some of us spent one more night there in our tents, some two, and then we quitted it finally.

What was accomplished during our journey of 1896 in some respects surpassed, in others fell short, of our expectations.

We crossed overland from Advent Bay to Klok Bay, from Klok Bay to Sassen Bay, and from Sassen to Agardh Bay, on the east coast, and back to Advent Bay. We made in all thirteen mountain ascents. We brought home a sketch survey of an area of about 600 square miles in the heart of the interesting middle belt of the country, besides a more rapid outline survey of the hills on either side of Wijde Bay. In addition, fortune enabled us to perform, round the coasts of Spitsbergen, the most complete voyage of reconnaissance ever accomplished in a single season. We almost circumnavigated the main island. We visited and entered to their heads all the great fjords that penetrate it, except Van Keulen's Bay, Cross Bay, Liefde Bay, and Lomme Bay. We saw the west, north, and south coasts of North-East Land from

Cape Platen round to Cape Mohn. We landed at the Seven Islands and closely approached Wiches Land (King Carl's Land). We brought back about 600 photographs of all parts of Spitsbergen. Such were our topographical results. The scientific results were more important, and will be duly chronicled hereafter. Our collections are in the National Museums at South Kensington and Kew, where they fill certain gaps—notably in the case of the geological collection.

We could have done more surveying had the weather been less persistently foggy. You cannot survey what you cannot see. With better sledges we might have covered more ground. As it was, we accomplished all that, I believe, would have been possible for any one to accomplish in the time and with the means at our disposal. I look back upon the season as one fruitfully and upon the whole pleasantly spent. The fogs condoned their sins against the plane-table with entrancing charms for the eye. The bogs are not miserable to memory. Sometimes the sun shone for days and nights together upon landscapes woven of sunlight and silver. Of such tapestries how can one's memory be dispossessed? Even had we accomplished no exploration nor added aught to scientific knowledge, the journey would have been worth while for the mere pleasure of it. That we may share this pleasure with a wider circle is the modest reason for the publication of the following narrative.

SKETCHING UNDER DIFFICULTIES.

CHAPTER II

LONDON TO THE ARCTIC CIRCLE

SELDOM for me did a London day open more peacefully than on the summer morning of June 2, 1896. The air was still. The sun shone softly through a light veil of mist, as it was destined once or twice to shine upon us by the shores of Ice Fjord. Thrushes sang in the mulberry tree over my breakfast-table. The sounds of London were faint and seemed remote. Every preparation was complete ; there was nothing to be done.

An hour later the five members of our united party were wrestling with forty-two pieces of baggage in the maze and scrimmage of King's Cross Railway Station. There was Gregory carrying a bundle of geological hammers and crow-bars, tied up with an old Snider, brown with East African rust. There was Trevor-Battye in a suit of clothes acquired in Moscow, when he arrived there in rags from Kolguev. Garwood with an armful of newspapers, and my cousin with an easel, poking out of a bundle of rugs, completed the party, to which many kind friends had come to wish good-bye and good luck. We missed only Mr. B. V. Darbishire, who was to have been our cartographer. Sudden and severe illness kept him back. A good German friend of his, with kind providence, had sent two Westphalian hams for consumption in Spitsbergen. "Where are my hams?" he pathetically asked, when I went to bid him farewell. "Safe in Trondhjem," was my scarcely regretful answer. Thus the savour of him went with us.

Before the train had gone twenty miles, the first accident occurred; a coal smut flung itself into Trevor-Battye's eye. "Shut the window, and blow your nose," was some one's advice. Battye followed it. "But why shut the window?" he said; "does that pull out the smut?" It was a hot journey, a kindly heat which gave us at all events one day's sense of summer. By five P.M. we were on board the s.s. *El Dorado* off Hull, and the screw began to turn. The air was again soft and damp; sea and sky to the eastward were green and grey, only separated from one another by a narrow broken line of shore and flat land, with here a row of trees and there a cottage. Now and again a barge came floating by, the only thing sharply defined in the midst of nebulous surroundings. Distance soon swallowed up the town, with its line of houses, its big church tower, its forest of masts, and its roof of smoke. The sea could scarcely have been calmer. The ship was not crowded. There was room for all, and no one was ill. Next day the same conditions were maintained, save that towards noon a little motion arose, and people grew somewhat silent, grave, and grey; but in the long-delayed evening perfect calm reigned once more; and when, about ten P.M. in the late twilight, we came into the sweet smell of the land, and amongst lagoon-like bays and low rock-islands, rounded by ancient ice, cheerfulness returned, and the light of expectancy was in every eye.

Stavanger lay almost asleep at the head of its green inlet. We landed there before midnight to see the exterior of its cathedral, which has been remade rather than restored, and to feel the joy of land beneath our feet. I for one was but half awake, and dimly remember the new cut tracery of the windows and carving of the pointed west door and Romanesque north porch—all nineteenth century alike. Our ship soon sailed again with a dormant company of passengers, who reappeared next morning (June 4th) to watch the passing of fine views, seen through a transparent veil of rain. At nine A.M. the pale daylight showed low islands, rounded and sometimes wooded, forming various combinations, and moving one behind another, the remoter seen over the heads of those nearer at hand. Snow-capped hills in the background looked large by contrast; and everywhere the smooth sea lay in the lap of the land, like clouds in mountain valleys. Wet on rocks and grass enforced their brilliancy of colour; but damp made the air grey, and almost as palpable as eider-down. The steamer pushed us as gently through it as it did through the water.

By ten o'clock we were in the Bergen Customs House, meeting with much civility from the officials, thanks no doubt to a good word from the Minister at the English Court. I inadvertently jeopardised the smoothness of relations by speaking of him as the Swedish Minister. "We have nothing to do with Sweden," said the official testily, "and no Swedish Minister has anything to do with us." To search for the *Raftsund*, the steamer that was to carry us north, was the first business. She was not alongside, as expected, nor even in sight. A lengthy investigation revealed her in dry dock, in the hands of shipwrights. For three days she would not be ready to sail.

More curious persons would have used the interval for an excursion; but when strenuous work lies ahead, delay produces a feeling of restless impatience that hinders the invention of immediate employments. Besides, I was cumbered with small anxieties about baggage and ponies, for, contrary to my orders, it had been arranged that ponies should be supplied to us here, instead of at Tromsö. I feared the beasts would be unsuitable, but could not see them, for they had not arrived, and only came just in time to be put on board. There are many worse places to idle in than Bergen, which has its fair share of sights, first among which is assuredly the Hanseatic Museum. It is an old Hanse merchant's house, kept in the state in which it was used, with its internal arrangements and furniture complete—the unwholesome cupboards for master, foreman, and apprentices to sleep in; the secret staircases, cunningly contrived for purposes of intrigue; the light weights for selling and the heavy weights for buying; the old ledger with its usurious entries; and "the key of the dairy" —an instrument of correction for refractory apprentices. The owner of the place, a local antiquary, humorous and original, congratulated us on the absence of ladies, and proceeded to explain its history and mysteries with many a giggled insinuation.

Next came the fish-market, with its tanks full of live fish, whose names and points the vendors described in language we could not understand, without ten words of Norwegian at our combined command. One large leaden-coloured fish, with a great oval head (a Gymnetrus, Norsk *Sildekonge*), in a tank full of cod, was shown off for our delectation. Its owner scooped it out, and, seizing it firmly round the throat, gave it a lump of wood to bite. The brute caught on like a bull-dog and rent the corner off. A more wicked mouth I never saw out of a criminal's head, and I believed immediately in the devil with a perfectly mediæval conviction. Old women came to cheapen cod, and plucked them about with horny fingers, standing by and laughing as knives were skilfully plunged into the heads of their selections and each throat was cut, and backbone broken near the tail.

One afternoon or evening some of us walked up the northern hill to the lower edge of the soft cloud-blanket, where a long traversing road, commanding wide views, led to a pretty restaurant. The meal was served on a terrace, whilst Norse airs were played by a rudimentary band. The evening seemed never to come, till we looked at a clock and found that the hour was midnight.

Another evening I climbed to the top of the Floi hill, and then wound away by a track leading to lonely uplands, remote from the very memory of man. Everywhere was the writing of the icy hand. All rocks were rounded. Thin carpets of earth or bog filled little hollows or held on to ledges. Patches of snow lingered in sheltered spots. From the top was a notable view all around, the pale sun setting amongst north-western clouds, and casting a glamour upon a network of sounds, where sea and land mingled in intricate interlocking, whilst promontories and islands became lower and more suave of outline as they lost themselves in the damp atmosphere of the mysterious sea.

A little village, immediately beneath, dipped its feet in the fjord, some of its houses rising plumb on all sides from the water. Ships of antique type, such as Vikings used, lay becalmed upon the near expanse, whose bright surface was broken by reticulating systems of ripples covering a wide area with formal decoration. Inland were blue hills, rising from a bluer and transparent sea, whose level surface was the top of a layer of smoke that had drifted from Bergen, and, by losing its grosser particles, had become thus endued with the rich glories of a counterfeit sky. Through the still air came no sound that the ear could distinguish, save the cry of a remote cuckoo and its fainter echo.

Another time I wandered round on the lower level to the reservoir lake, a natural basin enlarged by a dam, artificial but not ugly. A rock promontory juts into the water and offers a natural pedestal for a recumbent man. I lay there long in entire solitude with the black waters of the lake around me, and sparsely wooded and lonely hill slopes reaching up to a low grey roof of cloud. Cuckoo! cuckoo! was again nature's only song, with the faint lapping of water on my rock for its accompaniment.

At last, on the dull morning of the 7th, the *Raftsund* cast loose and steamed away through scenery that was at first tame and uninteresting, low-lying islands and rocky promontories all rounded in the same manner by the ancient ice-sheet. Then bolder outlines appeared, and seaward, abrupter humps and even steep-sided domes of rock, with now and then a snow patch. Inland came mountains of a certain size but lacking dignity of form. Near the mouth of the Nord Fjord the *Raftsund* followed a narrow channel immediately below Horneln, a bold and noble mountain that rises by steep slopes from the sea, and juts a bare precipice of nodding and splintered rock aloft to a jagged crest. The gloomy evening light and the grey cloud roof overhead, formed a suitable setting for this weird and solitary tower, which then looked as lofty and massive as hills and precipices can look, whatever may be their measured size. Clouds ultimately gathered about and hid Horneln, but in the other direction the sun shone forth and drew a clear fess of gold across the grey azure field of cloudy sky. The boat heaved beneath the ocean swell as she quitted the shelter of islands, and I knew no more till the morrow's awakening in the harbour of Christiansund.

Lovers of the sea and those to whom its motion is kind

can have no conception of the joy a true landsman feels whenever he quits, if but for an hour, one of those hateful prisons of the deep called ships. Words cannot describe my normal loathing for the sea, save as a floor to look down upon from a height of not less than 1000 feet, when the air endows it with an aspect of repose not its own. Then indeed it becomes glorious, and the ships upon it resemble fairy creatures, and the sun broods over it like a divine presence. My altar to Neptune shall be raised upon a hill, and thence will I offer him countless hecatombs. Such, however, is the innate folly of man, that, when he sees a beautiful view, he desires to be in the midst of it. "How fair it looks!" He thinks to gain by going where beauty seems to be. But the beauty is not there, but here, whence it is beheld. Not on that golden surface of the rippled sea, not on that rose-tinted peak, but here. Not in the remembered past, not in the brightly promising future, but, if anywhere, here and now. Tell a man this a thousand times; repeat it to yourself again and again. It is useless. Where beauty is seen, there would we be—thither will we—beside the great ones in history (who doubtless lived miserable lives), anywhere that looks fair; *dort wo Du nicht bist, dort ist das Glück.* But go there and you will find it flown, the glamour gone further on, or worse, further back.

Christiansund is doubtless commonplace enough. The point I thought I should always remember about it was the sense of solidity of the earth underfoot. Any road that led inland was good enough for us. We chose one that struck uphill toward a moutonnised moor. It ended in a picturesque wooden shed, a rope-walk that was tidied up and put away for the summer. Behind came the hill-top, and a little rocky park with a tower on its summit, to which we scrambled, going through bogs and up little walls of rock. A girl was selling pink fizzing drinks and biscuits. Gregory and I lunched on them, as we lay on the roof of the tower, with fjords and islands spreading around—blue hills, blue sky, and waters blue. It was a brief repose. Soon we were on board again, and our boat was struggling against a strong tide through a narrow gate. Rain drove us below to Garwood's dark-room, where experimental negatives were developed, and the usual faults of commercially made apparatus began to declare themselves.

Trondhjem was reached about midnight. Here the useless ponies were to be sold at as small a sacrifice as might be. Here too the heavy baggage, come through from London, was to be met and taken over. And here our sixth, and, as it proved, very temporary member, was to join us. Two hours finished the business, and left me free to visit the famous cathedral, once so beautiful, but now being supplanted by a modern copy, imitating the work of many ages and styles with a uniform finish and coldly accurate technique. Some remaining fragments of the original sculpture in the soft local saponite make the heart sad for what has been removed.

The old building, interesting and charming as the accretion of centuries, to Englishmen especially interesting for its evidence of English influence in Norman and early English times, has been treated as a model or mere design for a new one. It is a dreadful pity. If a copy had to be made one may admit that it is being well done. But who cares for copies. The ruggedest wreck of an original is worth them all put together. If a new building is required, brave men should venture on a new design, the outgrowth of themselves and their own day.

Gregory and I went forth by the road along the left bank of the Nid, vaguely aiming at the Lervoss. A mile outside the town a little stone-carrier's cart passed along, driven by a small boy. It had two small wheels in front and two larger behind, and on the axle of the larger was hinged a sideless floor, resting on the front axle, but ready to tip over backward at the smallest pressure behind. We called after the boy to stop and we scrambled on, all three of us crowding on a small sack and holding ourselves tightly together, only just keeping our elbows clear of the wheels. When the pony trotted we were jostled and jolted finely, and felt the livers within us. "Where is Lervoss?" we at last thought it time to ask. "Miles behind," was the answer. So we tipped the youngster and struck away at right angles from the road, heading for the river, which was somewhere in that direction. A field track led to a farm-house with a flying flag and a big swing in the garden. But no food was to be got there by hungry wayfarers, so we followed the track to another beflagged farm, where a crowd of men were gathered in their Sunday best. One spoke good American; he was home on a holiday from Minnesota. He made inquiry for us whether we could get a meal, and the hostess came forward and invited us in. "You seem very gay hereabouts with all these flags flying," we said. "Is it a wedding or a public holiday?" "It is neither," they replied, "but a funeral; this is the way we show our friendship for the dead." We talked with the American about the Presidential election. He was all for M'Kinley. The hard times in the United States dated for him from the Democratic Free Trade platform. "I'm a sound money man," he said. "I'm in favour of whatever I think likely to make trade look up." The meal was soon served. Thirty male guests sat down in one room; we had a table in another. The woman served the men and stood about waiting in our room, where a mother nursed her baby, which laughed and crowed at us in the intervals of its fare. The hostess came often to press her viands upon us with kind hospitality. A moderated gaiety pervaded the company.

Hunger being satisfied, we bade our friends adieu and sallied forth through an undulating and fertile country. Green fields decked the broken slopes. Approaching the dip to the river's trough we saw over a wide area, diversified by old moraines broken down by alluvial action, and terraces deposited at the opening of gorges. On the summit of a wooded knoll a roar of water saluted our ears, and we looked down upon the crest of the upper Voss, relieved against a sunlit cloud of water-dust flung into the air from the foot of the fall. Following a track down the water side we enjoyed the cascade from many points of view. Few pictures and no photographs give the least idea of the charm of any waterfall, for its glory is not in its form, but in its weight and volume, which the motion and roar of the water reveal. The river, that here tumbles over a cliff, was this day brimful, submerging its green banks; the brown smooth slope of the torrent at the top of the fall, hid within its depths volumes of air, swirled in by the first rush. These came to the surface like bursting shells a few yards down, and thenceforward the whole was a mass of cream-white foam, visibly heavy and of mighty volume. The setting sun shone broadly on the face of the fall, but left in darkness the curving reach of the wide river below. A path led through a wood, with open grassy places, in one of which was a family — parents, aunts, and children — playing kiss-in-the-ring, and looking like so many fauns, for which we blessed them. The second fall came into view round a corner, and the path led to a restaurant-platform. This fall is one undivided shoot of water, less impressive, we thought, than the upper cataract, perhaps because we saw that first. Waterfalls seldom gain on acquaintance—it is the first impression, the contrast of then mass and hurry and might with the stillness from which they emerge, that impresses the beholder. As the contrast is forgotten the effect diminishes. Always arrange to come suddenly on cataracts.

A road led from the restaurant. We elected to follow the river-bank, to which a track invited us. It ended in a swampy wood, through which we urged a laborious way, sinking ankle-deep in mud, whilst wet branches hit us on the face. Climbing a hill we came to drier places, then to a real path, and then a road. Five miles' plodding brought us to the ship about eleven o'clock in the so-called night.

Early one morning we steamed away to the neighbouring Orkedalsören to pick up the materials for the Spitsbergen inn. Here is one of the largest timber works of Norway, where the hillside harvests of the woods are gathered and wrought into every form. You can buy a ready-made house as easily here as a wooden trunk. The building was all ready, was in fact standing in the yard, its two storeys in separate places. To pull it to pieces, load it on trucks, and run them along the railway and the wooden pier to the steamer, was not half-a-day's work. In a few hours we sailed again.

The scenery continued dull—rounded rock-islands under a roof of fog, which hid all bolder prominences. But late in the evening it improved, and we had one lovely view when we gathered on deck in our warmest wraps. The cold air was utterly clear. A long mountain outline, finely complex, divided a purple range of hills, that looked flat in their remoteness like a wall, from a band of sky, yellow with radiance poured level from the lowering sun, whose wake upon the calm water of the sound was broken by the dark silhouette of an anchored schooner. Little, however, in a general way, cared we for weather or scenery, for there was enough to do with our things and with one another. I read

Gregory's "Rift Valley" with immense delight, and the additional advantage of being able to apply to him for a running commentary, whereby the *i*'s of the tale were merrily dotted and its *t*'s luminously crossed. I kept pressing him for hints about travellers' equipment, derived from his own actual experience, and jotted down in my notebook this advice: "Travellers should take with them a sausage machine for tough meat, a bellows for bad fuel, and a packet of seeds of quick-growing vegetables." Like most men he only partially practised his preaching, for there we were with neither sausage machine nor bellows. We found him a mine of valuable and peculiar information, most of which I have forgotten, though I carefully stowed away the fact (whose historical accuracy I have not verified) that Tristan d'Acunha is an island where one of the unappropriated female majority carried off and married a shipwrecked sailor before he had time to dry.

There is always work to be done to baggage by a fastidious traveller. Studley overhauled the reindeer-skin sleeping-bags, and improved upon his by sewing up both ends and opening a hole in the side. After three hours' stitching he made experiment of the result, and flopped about on the deck like a small walrus. "It'll be jolly hot," said some one. "But you'll want it to be jolly hot," replied the voice within the bag; "it'll be jolly cold, you bet." Meanwhile Garwood was vigorously testing our photographic outfit with the usual result; such is the phenomenal ineptitude of photographic manufacturers from a traveller's point of view. He found that the cut films were in many cases too long to go into the film-carriers and needed to have their ends reduced, a process of course impossible of performance within the narrow limits of a changing-bag, when fingers are blue with cold. He found that one of our changing-bags was so stupidly contrived that it could not be seen into through the glass made for that purpose. He found that a whole supply of

J. T. STUDLEY.

plates, advertised, bought, and specially labelled as thin, were of merely normal thickness, and would not go into the double-backs made for them. And finally he found that, in a new film roll-holder, the cylinder would not turn, and only rent in half the film it was intended to wind off.

A tremendous amount of sewing was accomplished by industrious persons. Gregory and Trevor-Battye "fixed up contraptions" for capturing minute marine organisms, and almost every one had something to set to rights. Bachelors came out strong with hus'ifs, which had all the appearance of being the handiwork of gracious maidens. One particnlarly favoured individual confessed to having received no less than seven as gifts. They were primed with needles, buttons, and the like, but no one had put in hooks and eyes (which we chanced to want), nor were there needles of triangular section, nor thimbles that would go on to thumbs. I observed that, after an hour's sewing, the hand of even the most skilful was bleeding from various wounds.

Trevor-Battye and Studley had deep consultation over a book of flies, made by the latter with admirable skill. Intricate was their talk and wide their reminiscence of fish killed in many lands—those grand days in each man's past that make up the sum of pleasant memory.

Gaily laughed the morning that followed; gaily danced the sunshine on the waters of successive sounds, with sky clear, air fresh and stimulating. Graceful black-backed gulls followed in our wake, and frequent cider-ducks and drakes flew aside. From hour to hour we passed islands and promontories, and threaded a way through broad sounds or narrow passages, sometimes emerging into the open sea, but only to find it as still and kindly as a lake. If ocean travelling were always like this, I could understand the attraction of a sailor's life. Yet the air was not still, for there was a merry breeze before which scudded double-prowed boats, assuredly the gracefullest craft in the world.

"Blue ran the flash across" the hills. Over the ship's side were crystal waters through whose shallower places the eye could search the floor of purple rocks and white sand.

The mountains grew bolder, the scenery nobler further north. The Seven White Sisters, first faintly seen ahead, waxed in majesty as they approached. From hour to hour more snow was in sight. Yet the view was never in the least like a view in the Alps, even allowing for the presence of the sea. Of characteristic grace are the domed islands, like pierced Torghatten, which rise from the water in gentlest sloping lines, then swiftly lift themselves aloft with ever steepening curve to a broad summit. Some mighty masses, like the Lion Island, front the sea with imposing precipices. Distances became deceptive in the increasing clearness. We mistook a gully filled with a lofty waterfall for a snow couloir. The angular velocity of the falling water was so small that the eye overlooked the motion. Thus we had visible proof of the fact that the charm and strong impression which cascades produce upon the mind is due, not to the form (save in exceptional cases, like Tosa Falls or the Seven Sisters in the Geiranger Fjord), but to the motion and perceived weight and volume of the falling water.

In the evening all came on deck to "dilate with the right emotion" while crossing the Arctic Circle. Nature has marked the position of this mystic line by an island of notable form, named Hestmand, which rises boldly from the sea and culminates in a jutting finger of rock. On all hands the scenery was divine in the pellucid air. Thirty miles away to the west, far out in the ocean, the storm-rent Traenen Islands lay like a cluster of opals. The mountains everywhere took bolder forms, with weathered precipices, snow outlined ledges, and toothed arêtes. Dancing light made patterns in the water, leaving polished highways and spaces patched about. The ship swung through narrow passages, and made constant changes of view and ever fresh combinations of noble forms. Fjords opened inland, but only to close again with fascinating rapidity. The air was cold, the sun bright ; it was a joy to live. That night the sun did not set ; it went below a bank of soft grey mist, and for a while was not seen, but it remained above the horizon then and thenceforward. We were north of the midsummer night.

UNPACKING.

CHAPTER III

TO SPITSBERGEN

ON awakening next day (June 13) and looking out through the port-hole, a superb sight met my gaze—the long white front of the Lofoten Islands, stretching away beyond the blue sea into pale distance, Vaaga-kallen, a bold pyramidal peak, the centre and culmination of the range. This glorious scene ushered in a day of perfect delight. Hastening on deck in pyjamas I proceeded, in the sheer excitement of joy, to ruin a series of photograph plates—exposing one twice, the next not at all, breaking a third, taking the double-back off with a fourth before closing the shutter—and so on, the result being that the view was hidden without being photographed at all. By this time the *Raftsund* had entered the charming harbour of

Svolvaer, enclosed and backed by glorious snowy mountains. "What a place," we all cried, "for a summer holiday!" and we registered vows to come thither again. But it is really in March that one should visit Svolvaer, when the great fishing takes place, and thousands, tens of thousands, of the beautiful boats of the Nordland are gathered together at one time for the fishery. One moment they will all be busy with the catch; the next a breeze may spring up or the fish move away, and with a single impulse ten thousand sails will be hoisted, all bright with colour, and the fleet will move off, urged as by a single will.

The studio and pictures of the late talented young Norwegian artist, Gunner Berg, attracted us ashore. The streets of the village are creeks and its blocks islands. A boatman rowed us to the house we wanted, and fetched Gunner Berg's brother, who kindly gave admission to the studio. It contains little but oil sketches and unfinished pictures. The Vaagakallen peak in deep winter snow, with groups of bright-coloured fishing-boats on the waters at its foot, were the subject of many. A sketch of a skilöbner specially attracted me. The art of the painter was undoubtedly great. He depicted his own folk in the atmosphere in which they live. His work is full of air, light, and motion. He painted less the shapes than the relations of things. His sketches are not mere compilations from scenes beheld, but skilful renderings of definite impressions seized, comprehended, and made visible for others.

Shortly after noon we sailed again and looked up the Ohel Sound, to its continuation, the Raftsund, with Mö Sadlen at the end, biggest of peaks hereabouts. Fresh water was taken in at picturesque Brettesnaes, which boasts an English vice-consul (who came to call), and a manufactory where cods' heads are converted into manure. Her thirst quenched, the ship sailed on, past the glorious

Troldtind, from whose main backbone buttress ridges descend, each rising to a mighty tower before sinking to its foot, feathery with young green birch, by the margin of the deep blue fjord. These towers are truly Alpine, almost Himalayan, in style though not in scale. The snow slopes and couloirs upon them were striped with the tracks of many falling stones. Valleys, white to the water's edge with deep winter snow-beds, intervened between the buttress ridges. Each opened in turn as we passed, till one came deeper and between walls bolder and loftier than the rest. Its floor lies far beneath the waters, which, flowing in, form within the heart of the mountains the famous Trold Fjord. Mr. With, manager of the Vesteraalen Company, who has a keen eye for beauty and delights to share the joy of it with his companions, caused the steamer to enter this beautiful place for our delight. The spring snow, lying low down to the water's edge, added, I fancy, to the dignity of the scene, for the sun shone brightly upon it, so that the end of the fjord, and the cirque of mountains closing it in, looked like some giant palace of silver seen through the narrow entrance, with its mighty door-posts of dark precipitous rocks. The end of the fjord was a pool, so calm in the bright day, with snow gullies and waterfalls descending to it, each in its place mirrored below. The ship turned slowly round, almost grazing the banks, and steamed out again, leaving this sanctuary of Nature for a while polluted with our smoke. How poor a return man makes to Nature for her gifts!

Advancing up the Raftsund, cirques and precipices succeeded one another, and couloirs and snow-slopes seamed by falling stones. At the end of the sound came a large, open area of water, whence many smaller sounds radiate. Brilliant it was, with the water so still, the air so bright, and the fine ring of mountains all around. We were to behold them presently from another standpoint, and in still greater splendour. A twelve hours' halt was made at Stokmarknaes, to take on coal and give Mr. With time to visit his home. A hill-climber himself, he knew the best point of view, so at seven in the evening, Gregory and I set off to follow his advice, and climb the round summit of the fjeld behind the town. Studley and Garwood went fishing and caught eleven sizable cod, whilst Trevor-Battye and our artist sketched the range of snowy mountains across the fjord.

Gregory went ahead like a steam-engine, whilst I did the puffing and blowing behind. An hour's tramp carried us over fields and bogs, through a wood of stunted birch, and up rather steep snow-beds and a rounded ridge, to the shoulder of our hill. There was a springy cushion of cloud-berry plants and a splendid view. On either hand were hollow places, once filled with glaciers, which, in some hurry of retreat, left their lateral, medial, and terminal moraines *in situ*, and disappeared. A stone avalanche, falling not far away, filled the quiet air with the echoes of its rattling. Half-an-hour's further trudge carried us up the remainder of the ridge and over the broad area of the culminating snow-field, along which the sun drew out our blue shadows to a hundred yards. I can recall no more enjoyable hour of active life than the one we spent on the upper part of this hill. On one side was a large snow-cornice overhanging a precipice. Elsewhere the spotless dome curved gradually away, and formed a smooth fore-ground of warm creamy white to what was assuredly one of the finest panoramas I ever saw. Golden light flooded forth from the low sun, and enveloped the north in a glory as of the portals of heaven. A broad water-highway led to the sea. Along its far side ran the straight range of beautiful snow mountains of Lofoten, with splintered peaks rising from rounded slopes, which dipped under a flat alluvial foot. Where the peaks came nearest to us they were rich in colour, and clear in outline and mass, but they stretched away and yet away into fainter and tenderer delicacy of light and tone, till in the remote distance they seemed more impalpable than clouds floating in the rarest air. In other directions great reaches of water led the eye to distances so great and clear that it was as if we beheld all the area of the north at one sweep, with countless mountains, promontories, and secret places, all apparently aloof from and forgotten of man. The entire scene was absolutely superb, perfectly harmonious in forms, in lights, and in opalescent colour, pervaded by an unruffled serenity. I turned round and round in silence and enchantment. The Arctic fever seized me at that moment and thrilled through every fibre.

Leisurely we descended eastward by another route to a second shoulder, where another halt was made, and I lay a-dreaming on a large flat rock, whilst Gregory, more industrious, sketched. Setting forward again towards the steep face, we started down a snow-slope, hoping for a glissade. But it was too hard and steep. Steps had to be carefully trodden, and the direction constantly changed to avoid places where ice came too near the surface. The slope contracted to a couloir. We took to its grassy side, but the ground beneath the thin vegetation was frozen hard and was almost as slippery as naked ice. There was no axe with us, so care was necessary. We let ourselves down by occasional birch bushes, and thus, sometimes taking to the couloir itself, gained the snow avalanche fan below and glissaded to its foot. A beautiful little lake, framed in green, and reflecting the low sunlight, here occupies the position of the foot of the ancient ice-fall that descended this corrie. Leaping fish made rings in the calm water. We scrambled round, and floundered across bogs and streams to the edge of the village, entering it through the main street of a quarter consisting of empty wooden houses and booths, occupied for a few days every June by the 10,000 people who frequent the great fair of Stokmarknaes. The ship was regained at eleven P.M., and such was the enthusiasm of our praise that Trevor-Battye and our artist were fired to emulation, and promptly set forth to follow our steps. They climbed the mountain to its top, but the colour was gone. The seemingly changeless vision we beheld had vanished like the dream of a moment that it was.

A pall of grey cloud settled down overhead in the early morning when we sailed again, painting the landscape in mere black and white, save where sometimes a remote hill shone pale yellow in a small island of transient sunlight. All day long bleak scenery defiled past the ship, enlivened now and again by a picturesque sail or a peak more individual and abrupt in form than its countless fellows. Cold breezes and showers drove us to seek employment below. I read in Gregory's book and found the following description of a happy African day: "In company with the Goanese commandant, I spent a pleasant afternoon catching lizards and scorpions, and digging up skulls." I feared Spitsbergen might seem dull, if that was his idea of bliss. Being a person of varied resources, however, he was actively employed measuring the details of four hundred specimens of a bone from the head of cod-fish. He said the pastime was excellent. A momentary excitement was caused by our touching bottom in the shallow channel of Risösund. The face of a cliff, overlooking a narrow channel between Bjerkö and Helö, is the resting-place of countless kittiwakes, who cumbered themselves little about our close proximity, though we blew the whistle a few yards from their nests, causing our own insides to quake like metallic resonators. In the late evening Tromsö was approached through a veil of rain.

Under ordinary circumstances a day in Tromsö would not be a wildly exciting experience. For us, however, it was busy enough. We had a horse-dealing transaction to begin with, resulting in the purchase of two ponies warranted to eat dried fish-heads or any other garbage, and that proved entirely satisfactory. Then there were our men to be interviewed—Pedersen the senior, an old sea-dog reputed to be

well versed in Spitsbergenography, and a strong man of less precisely defined qualifications, whose name was the Norwegian equivalent of Williamson. There was also business to be transacted with the vice-consul and at the post-office, further supplies of food to be bought, our whale-boat and its fittings to be inspected, the ponies to be shod, and a thousand and one more details to be finished out of hand. Incidentally we dropped into various little shops and inspected their stores of furs, horns, and the like oddments—for the most part costly and unattractive. One store was devoted to Lapp manufactures, described by a printed advertisement as "very seeworthy." A poor little Lapp came in whilst we were there, wearing leather clothes and turned-up shoes like a Hittite. I felt as though I had tumbled into the presence of a neolithic man, so broad seemed the tide of centuries audibly murmuring between us.

The ship, as the day advanced, became a mere pandemonium. Two or three friends of Mr. With came by invitation to make with him the trip to Spitsbergen and back—one of them was Mr. Ekroll of Lofoten, who had spent a winter on Edge's Land a year and a half before. There were also a French gentleman and a Stockholm journalist, Herr Stadling, going to join Herr Andrée. There were several boats and our ponies to be taken on board. There was a crow's-nest to be fitted up. All sorts of people came to say good-bye—consuls, agents, dealers, and relatives or acquaintances of passengers. The deck was crowded. Glasses clinked. Every one was in the way of every one else, but the sun shone brightly and all were gay. In the midst of the shindy, Gregory digested geological papers from various journals, cross-questioned any one that came handy about Spitsbergen birds or the Norwegian vocabulary, and went on piling up information generally. "You read always," said the French gentleman to him. "Yes," was the merry reply; "you see I am young and have a lot to learn." Garwood was also busy. He started from England with some work on hand still requiring a preface. With this he had been labouring in the intervals of sea-sickness, negative developing, baggage overhauling, and the rest, at odd moments during the voyage. The hour appointed for sailing came, but the preface lingered. Furiously the author worked on amongst teacups and the débris of food. At last the work was done. A wild rush on shore, and the precious document was consigned to the post and all the printer's devils.

At last it became clear that the ship would soon be ready to sail. Various Jeremiahs bade us good-bye, and expressed hopes for our success, which at the same time they prophesied would not be realised. They said the season was a late one, that the ice would keep us out of Ice Fjord, that we should not reach Advent Bay before August. Few were the prophets of smooth things. They derided our ponies. They knew of better men—as indeed they easily might—than those we had engaged. They wished, one and all, that we had come to them for advice, when we might have started with some chance of success. Thus we quitted Tromsö in a chastened mood, which, about the supper table, gave place to wildest merriment, for at last we felt that the preliminary stage of our journey was ending and that the fun would soon begin.

There remained, even yet, a final incident, when we unexpectedly stopped at Skaarö, close to the entrance into the northern ocean, and visited the whaling station there, about midnight. Three newly captured finner whales lay stranded or moored by the shore, and were in process of being flensed. One was a great fellow; the stink that arose from all was overwhelmingly horrible! The wife of the manager of the company invited us to her charming house, where success was pledged to our enterprise in champagne. A wood fire blazed on a picturesque hearth; the walls were hung with bearskins, bold autographs of Kaiser Wilhelm and other distinguished visitors, and various trophies. The windows commanded views over a landlocked sound to a row of the snowy mountains we had just come by. The cosiness and comfort of this hospitable home were most enjoyable.

A few minutes later we were again upon the ship. A brief passage carried us into the open sea, and our course was finally set for Spitsbergen.

Of what happened during the next twenty-four hours I have but the vaguest idea. The sea was what its admirers might call calm. There were no white caps on it, but a long rolling swell came from the south-west, enough to keep one

A WHALING ESTABLISHMENT.

miserable, even without the marrow-freezing wind. So I remained in my bunk and read, till the boredom of inactivity became intolerable. Meanwhile exterior surroundings were, I believe, interesting enough. Whales spouted, and two finners came within a harpoon-throw of the ship's side. The proper birds appeared in due succession. Little auks scurried about in flocks, or flapped along the surface of the water, and then dived beneath it in a hurry; fulmar petrels flew gracefully around and scudded along incredibly close to the water with apparently motionless wings. There were Mandt's guillemots, Brünnich's guillemots, black guillemots, kittiwakes, Arctic terns, Arctic puffins, and a pair of Pomatorrhine skuas, all duly and immediately recognised by Trevor-Battye with a readiness that seemed magical to me, who distinguish with difficulty between a thrush and a blackbird.

About two A.M. on the 17th, Bear Island was sighted to the eastward, but all we saw of it was the lower part of its northern extremity between a floor of grey sea and a low roof of grey cloud. We held steadily on our way, hour after hour, through the cold miserable air, and over the more miserable heaving sea. Shortly after noon the first fragments of ice were passed, while along the eastern horizon hung a curtain of the coldest white conceivable, like sunlight on a cloud seen through a veil of mist. It was an Ice Blink, sign of the presence of an ice-pack in that direction. An hour later we were in the middle of drift ice, and had to slow down and wind about to avoid the pieces. These were of all shapes and sizes, distributed with average distance between them of perhaps a hundred yards or less. On the horizon they appeared like breakers. Nearer at hand they took the queerest shapes, like swans, rearing horses, people seated at table, camels, lions, buildings, what-you-please. Often there was a covering of old white snow on the top; then a thickness cut through or undermined by the wash of the waves, or hollowed out into caves. This part was always divinely blue, with a blueness more delicate and pure than that of an Alpine glacier's crevasse. The mass below the water had a flat upper surface, and was much broader and more solid than the part above. It showed a light bluish green tint through the water. The waves breaking over this greenish base swished and flopped under the overhanging masses, and no doubt were rapidly destroying them. Some larger fragments of dirty land-ice intruded themselves amongst these delicate creations of the bays or the open sea.

A ship intended to face ice is always built of wood, and, if a steamer, furnished with a two-bladed propeller. Our vessel was of iron, and her propeller had more blades than two. It was therefore absolutely essential to avoid coming in contact with the drift ice. We wound about amongst the masses, but the further we advanced the more numerous they became, so that we were at last compelled to steam south-west for an open place visible from the crow's-nest. Before reaching it a band of larger ice-masses had to be passed, floes a quarter of an acre in area or more, which had been crushed up, one upon another, and piled into strangely contorted forms. On one of these floes lay half-a-dozen young saddle-back seals, which let us come so near that we were about to launch a boat and attempt to harpoon them, but they awoke, slipped off into the water, and were gone. Plenty more seals appeared during the day. They often held themselves up high out of the water, hideously humanly, and gazed at the ship, before diving and hurrying away.

However interesting our surroundings may have been, and they were sometimes even exciting, what all were pre-eminently conscious of was the cold. Passing strange were the costumes adopted. One and another would vanish below and return with some additional wrap. Studley became quite stylish in corduroy breeches of the most fashionable hunting cut, wherewith his long rubber boots, stuffed out with hay, by no means harmonised. Trevor-Battye produced Samoyede pimmies. Coats of leather, furs, and every kind of thick outer covering appeared; but the greatest variety was in caps, which ranged from a thick Kashmiri turban through every grade of fur head-dress to the ordinary sou'-wester. In the baggage were four pairs of enormous fur-lined boots brought by mistake, and which we intended to return unused. They were all appropriated early in the day with thankfulness, and we slid and waddled about in them, awkward but comfortable.

When attention was called to the fact that this 17th of June

DRIFT ICE OFF SPITSBERGEN.

1896 was the three-hundredth anniversary of the discovery of Spitsbergen by Barendsz, the ship was gaily dressed with bunting. Fog threatened to prevent us from completing the celebration by beholding the island itself, but about eight P.M. the fog lifted, and we saw, some twelve miles or more away to the ESE., the bases of the hills about Horn Sound and part of Mount Hedgehog, the whole deeply clothed in snow, creamy white in the sunshine, and faint blue in the shadows, but too distant to be picturesque or at all striking. Ice thickly fringed the coast, and Horn Sound was seen from the crow's-nest to be blocked with it. Slowly we crept along up the coast through the scattered drift ice, sometimes forced to move farther out to sea, but always trying to edge back towards land as soon as possible.

By midnight the sea was absolutely and divinely calm, with the gentlest heave coming across it, a bright grey flat, silver grey under a lighter grey sky. Scattered about in it were the loveliest masses of ice, blue, indescribably blue below, with caverns of darker blue and white surfaces above —things fairy-like, bathed in the soft grey air, melting away in the distance, yet strangely defined and clear, with a soft definition and a clearness as of gossamer fabrics. Landward was just a stripe of faint yellow, where the snowy foot of hills appeared, illumined by the pale midnight sun.

Next morning when we came on deck there was a great transformation, for the day was absolutely clear, and we were anchored off Cape Starashchin at the mouth of Ice Fjord. The sight that met our eyes was a vision of beauty so radiant and glorious as to seem past the possible perfection of this world. Turquoise blue was the sea, dotted about with white ice-masses, each a thing of beauty. Snowy hills framed the water, white hills with sharp rock arêtes and boldly-bedded slopes. Southward the Russian Valley penetrated the land, divided by the blade-like mass of Mount Starashchin from the sea, and by a carboniferous ridge on the other side from Green Harbour. North was the fine mass of Dead Man's Mount, with the fainter hills and headlands of the Foreland further out. Looking up the fjord were the coasts and bays we were soon to know so well; the glacier fronts near Cape Boheman, the white hills on the south beyond Green Harbour, remains of a broken-down plateau now cut up by deep valleys, all white with winter snow. The place where we were anchored was close to the southern shore, near a solitary rock named the Fortress, just off the mouth of the Russian Valley. Half a century ago this was the site of a large settlement of Russian trappers. One of them, by name Starashchin, spent, it is said, no less than thirty-seven winters here. Some say that he died in Russia, others point out his grave near the Fortress. The man is often mentioned by old visitors to Ice Fjord, but little seems to be recorded about him except the length of his stay. He used to return to Russia for the summer.

At an early hour we set forth to land. Rowing against the wind infused warmth into us. We wound about among stranded masses of ice, hollowed below by the sea, but with smooth beds of snow above. Sooner or later the stalk of ice supporting the snow table becomes slender and snaps, or the balance of the whole mass is disturbed and it tips over on its side so that the table slopes up out of the water. This is the fundamental type from which are developed all those queer shapes, as of bird and beast, with which drift ice diversifies the sea. Birds flew close to us or dived into the water by our side—king eiders and common eiders, little auks, puffins, and guillemots. Bravest of all was the fulmar petrel, which hovered close over our heads, then swung away with easy grace, only to circle round again and return to continue observations on the intruders.

Snow lay deep by the water's edge, and overhung the shore with a splendid white cornice. Landing at a suitable gap, we climbed on to the flat—a raised beach of

FLOATING ICE.

THE NORTH COAST OF ICE FJORD, AT THE ENTRANCE.

water-rolled pebbles covered in places with springy vegetation, and already flushed with pink saxifrages (Oppositifolia) lovely to behold. Gregory went to work, smashing things with his hammer and filling his vasculum with plants. We tramped across the flat, for the most part over snow, and climbed a slope eastward to the lowest crest of the long ridge that forms the east bank of the Russian valley. From the top a view over the whole of Ice Fjord burst upon our delighted gaze. Green Harbour's mouth was close at hand with only broken ice before it, but beyond came the pack blocking Advent Bay and all the remoter fjords. Sassen Bay, Klaas Billen Bay, Dickson's Sound, Safe Haven—we could see them all, and brilliant mountains behind them, here peaked and beridged, there rising as by steps to broad flat levels, but all alike—peaks and valleys—bright with a deep covering of purest snow. This splendid ring of white enframed the great fjord, which too was for the most part white, but blue near at hand, dotted over with ice-blocks, and so calm that the hills were reflected in it, and lay amongst the floes.

From this point of view the ice-master gained the information he needed about the state of the ice and was able to make his plans. He unfortunately, with a sailor's strange secretiveness, kept them to himself, so that we were unable to use the interval to best advantage. While Gregory and I were on the carboniferous ridge, Trevor-Battye searched the level shore and swamps within for birds, and Ted sketched the snow-cornice overhanging the sea. A gay breeze carried us back to the ship with birds flying round and little waves following and laughing as they fell back and gave up the race. Garwood did not come on shore with the first boat but followed a little later. What he was doing no one knew, nor where he was, but he shall tell the story of his adventures in his own words.

THE ASCENT OF MOUNT STARASHCHIN

BY E. J. GARWOOD

Having landed near Cape Starashchin, after Conway and Gregory had gone away with the skipper to inspect the condition of the ice, I struck inland across the low swampy ground, which here extends for several miles between the hills and the sea, and directed my steps to the foot of a prominent mountain ridge running southward and nearly parallel to the seaboard. My original intention was to collect specimens of the rock, but habit is strong, and I had not gone far before the temptation to attempt the ascent of the peak overcame me. After examining the ridge for a practicable route, I changed my course, so as to arrive at a point more directly under the summit of the mountain, and so avoid traversing the whole length of the arête. This involved crossing the boggiest portion of the foreshore, and my first attempt at Arctic exploration was decidedly damping.

The low ground, which must be crossed, forms a portion of the recently-raised sea floor, which makes so conspicuous a fringe to all the more sheltered inlets on both sides of the island. Being composed of glacial débris deposited as a kind of submarine boulder clay, its consistency is loose and treacherous. Though for the most part still covered with snow this was melting so rapidly, now that the long Arctic day had set in, that the whole surface had been converted into a species of snow bog. During the six weeks which followed we had ample opportunity of studying the peculiar characteristics of these bogs, half snow, half water, and wholly abominable, but this walk, being my first experience of them, made perhaps the greatest impression upon me.

Here and there low domes of mud, from which the snow had melted, afforded a firmer resting-place for the foot. The formation of these protuberances by the swelling of the saturated ground under the expanding action of frost, and the consequent formation of shrinkage cracks, approximately hexagonal in shape, during the dry season, is a very characteristic feature of these lowland flats. Progress over this kind of ground was nearly as irritating as the obstacle race over loose moraine by candle-light, so frequently involved by an early start in the Alps. As I plunged up to the knees in the liquid snow bogs, splashed to the eyes, I caught myself quoting remarks which I remember hearing dropped by the man furthest removed from the candle on one of the above occasions. After nearly an hour and a half's tramping I gained the scree slopes of the mountain, and commenced a diagonal ascent to the foot of a projecting rib. This west side of the mountain is decidedly steep, and the nearly vertical buttresses are intersected by steep gullies, at that season still filled with snow.

Never did I come across a mountain in such a terrible state of repair; step after step gave way, and I do not think that during the whole of the ascent to the arête, a single hand or foot hold could be called really safe. Having started from the ship without any intention of climbing, I was without an ice-axe, and, moreover, hampered by a gun, a camera, and a geological hammer, but I abandoned the former soon after commencing the ascent. After mounting a short distance I was on the point of crossing the couloir to my right in hopes of finding firmer rock on the buttress beyond, when a portion of the cornice above broke off, and, gathering material from the upper part of the gully into which it fell, rushed past me down the narrow sinuous couloir, hissing and writhing like a serpent.

After this little exhibition of temper, I decided to stick to my rotten buttress, and, after removing a large portion of the mountain in my struggles, reached a small cornice which projected from the west side of the ridge. Cutting through this with the geological hammer, I gained the arête. The structure of the mountain now became clear, and I was no longer at a loss to account for its disintegrated condition. I was standing on the upturned edge of one of the harder grit bands, here interbedded with slates constituting a part of the Hecla Hook formation—a rock series apparently older than any of the fossiliferous strata on the island. In no place in the world can the disintegrating effects of frost be so admirably studied as in these latitudes: the copious discharge of water from the rapidly melting snow, during the continuous Arctic day, permeates all the cracks and saturates the rocks with water, which on the first frosty night is expanded into solid ice. In the case of the ridge of Starashchin, the strata having been tilted into a vertical position, access for the water is easily obtained along the truncated edges of the numerous bedding planes, causing slice after slice of the face to be wedged off, and shattered into incoherent piles of rock.

But my attention was soon diverted from the rocks at my feet to the magnificent panorama which the ridge

RUSSIAN VALLEY AND MOUNT STARASHCHIN.

commanded. I cannot hope to convey in words the beauty and grandeur of that view; the poet to the expedition might perhaps have done justice to it, but unfortunately I had left him shooting sandpiper on the marshes below. It was so similar and yet so different from the views to be seen on a fine day from any of the famous summits in the Alps.

It was my first peep into the scenery of Arctic lands. When coasting northwards on the previous day, thick banks of fog had hung in curtain-like folds along the land, increasing our curiosity in the country that lay behind. Now these had all rolled away, revealing a fairyland of ice and snow. The sun shone with a tempered glow in a wonderful sky of turquoise blue, a sky whose colour was different from anything I had ever seen above the snow-fields of the Alps, where, on cloudless days, owing to the absence of suspended particles, the colour of the sky often approaches black. My immediate interest lay in the direction of the interior, and I eagerly scanned the scene of our future operations.

Inland to the east and south the eye wandered over seemingly endless ranges of undulating snow: a few rocky peaks were beginning to push their dark points through the thick mantle of snow, accumulated during the last long Arctic night, like the first young shoots of the snowdrop on the approach of spring. Snow filled the valleys to the east in the direction of Coles Bay, damming back the drainage and forming lakes, and stretching shorewards till it merged into the frozen margin of the fjord. In the bay at my feet gigantic icebergs of a wondrous blue shimmered in the frosty light as they glided seawards on the ebbing tide. Beyond lay the ice-pack, and at the back of beyond lay that mysterious region whose secret so many had tried in vain to solve, and which, in spite of many an heroic effort, it still clasps tightly in its icy grasp. I thought of Nansen, that gallant Norseman, who, sailing northwards now three years ago, had drifted into the silence of that frozen waste; and, as I gazed, there crept over me a deep mysterious awe, a shadow from the threshold of the great unknown. It was a scene which I shall never forget.

But a long distance still separated me from the summit, and at any moment the captain might return and sound the signal for my recall, so stepping carefully along the broken arête, sometimes of ice, sometimes snow, I hurried as fast as possible to the foot of the first of the twin summits.

No difficulty occurred which would have caused trouble to a properly-constituted party, but climbing alone on a corniced ridge, without rope or ice-axe, was rather ticklish work; the situation, however, was not devoid of humour, and I laughed aloud, whilst cutting a staircase with my hammer down a nasty dip in the arête, about as wide as my boot, when I thought of the expression which the face of my old tutor, Joseph Imboden, would have worn if he could suddenly have come across me at that moment.

After many ups and downs, however, I arrived near the summit of the north peak; skirting below this, I kicked a passage across the snow face and rejoined the arête to the south, and was pounding along this, to what I considered must be the highest point of the ridge, when I thought I heard a faint whistle. Glancing towards the entrance of the bay, I could see a tiny puff of steam floating away from the funnel of the little toy steamer, and as I watched, there came a second little puff, followed by a faint whistle; this was the signal agreed upon for my recall. Glancing hurriedly round me, I exposed one film and then turned and fled.

For a time I kept along my previous track, but on reaching the foot of a long rise in the arête, I suddenly discovered that the steamer had left her anchorage and was steaming away up the fjord. Abandoning the arête, I turned straight down the west face of the mountain, thinking it better to risk a precipitous but rapid descent than run the chance of a lingering death from starvation.

My troubles soon began: the couloir down which I started became rapidly steeper; the snow into which I plunged up to my knees gave no support, and showed an evident intention of breaking away; and a little farther down, the sudden acceleration of a leg which I had tentatively advanced, showed the presence that I had dreaded of underlying ice. There was no choice but to continue the descent by the rotten rock ridge on my right. This involved great delay, armfuls of débris had to be pushed away, and a little platform constructed for each step: even then there was no feeling of security.

Suddenly the buttress I was descending stopped short at the junction of two couloirs. Like those farther to the north, which I had avoided in the morning, they were swept by avalanches from the arête above, and presented a steep surface of treacherous ice. Retreat was impossible, so, casting an apprehensive glance up the wicked-looking gully above me, I began rapidly cutting steps with my hammer across the narrowest part of the couloir. I fear these steps would not have passed muster by the editor of the Badminton book on Mountaineering, where we read that "the greatest number" of strokes "is required in cutting steps for a traverse of a very steep, ice-filled gully," and further, that "a good guide has been known to take seventy strokes to fashion a step"; but I must own that my dominating impulse was to reach the far side of the couloir as quickly as possible. Once, when nearly across, a stone tobogganed gracefully past me, serving, if possible, to hasten my movements. But at last I was across, and after this the slope lessened, and in another twenty minutes I had reached the scree slopes below. For the first time since leaving the arête I had leisure to look for the steamer: she was nowhere to be seen!

As I replaced my glasses after a fruitless examination of

with that abominable gully. However, the disablement of one of my cameras at the very start was too serious a loss to be accepted without a struggle, so, toiling up the rotten buttress once more, I managed to regain possession of the box. In returning, however, I dropped my fur cap while stooping to improve a step. It disappeared down the couloir, and there it lies, for a careful search at the bottom failed to reveal a trace of it.

As I jumped and waded along the five miles of swamp to the coast, I recollected that I had eaten nothing since breakfast on the ship, and it was already seven o'clock. The reflection was not a cheering one. The steamer had gone, and Battye and the artist had, I knew, returned long since to the ship. I glanced about for a sleeping-place and for something to shoot, and thought of the stories of marooned mariners. Only the night before Battye had recited to us a ballad of his own composition anent a marooned whaler, whose brain gave way under the strain of Arctic solitude. Some of the verses recurred to me, and seemed to describe very closely my own predicament.

"And who shall win when the fates begin to rustle their pinions black?
For the bergs that ride with wind and tide had driven the vessel back,
So that she lay ten miles away, low in a red sun's track.

This was the thing which, wearying in hunger, and alone,
Allan learned as he returned to drop on a barren stone,
Sick with the sense of his impotence, and with doubt of the drear unknown."

Nor was the sequel which describes the finding of the marooned man any more cheering. It ran something like this :—

"And out of the ground a figure wound through the roof of a lair of snow,
Weird as the theme of a graveyard dream, gaunt as a gallows crow,
And rocked itself on an icy shelf, moaningly and slow.

It sucked at the heel of a dead grey seal, like some wild creature caged,
And peered at the prize with puckered eyes, critical and aged,
Glancing askew at the presence new, as jealously enraged."

With these rhymes running in my head I reached our landing-place on the coast, and there, to my great relief, found our boat, and the artist still sketching. It appears the boat had been sent to wait for me, and the artist had taken the opportunity of returning to finish his picture. My first inquiry was for food, but there was none to be had, the party having returned to the ship to dine; but I espied a half-empty mug of beer, which had been standing in the sun since the morning; this I finished without ceremony and with great satisfaction. After an hour's pull, of which I shirked most of my share, we reached the ship, and I did ample justice to a cold supper, having been nearly twelve hours without food.

CHAPTER IV

ICE FJORD

WHILE Garwood was climbing Mount Starashchin, the *Raftsund* was trying to find a way through the ice towards Advent Bay. Fortunately for him she failed, and had to return. I was eager to land and explore the Russian Valley, but Gregory quite rightly protested. It was impossible to know when the ice might open, and to begin our explorations by being sundered into two parties forty miles away from one another, would doubtless have been unwise. So we spent the evening together on board, and Garwood presently returned and made our number complete.

As the sun went to the north a mellower beauty spread over the view, whilst a thin blanket of checkered clouds came up from the west. All day long, masses of ice had been drifting out of the fjord in endless procession, till it seemed as though there must come an exhaustion of the supply; but no diminution was visible, and no lead opened broad enough for the steamer to venture in. At any moment the slow-moving ice might change its direction and begin packing up at any point, so that without very great risk the *Raftsund* could not advance.

Early next morning (June 19) we were under way, at first through open water, then among larger floes. Off Coles Bay it was impossible to proceed; we had to return crestfallen a second time to the sheltered anchorage; whilst to make matters worse, the sky clouded over and the day became wretched. All the glory went from the view, and the distances were swallowed up in fog. How long we might be kept it was impossible to tell. We risked being left behind, and landed for a tramp over bogs and water-patches, covered with snow that let us through, knee-deep in freezing slush. Search for birds, flowers, and stones enlivened the otherwise dreary way to the mouth of the

ENTERING ICE FJORD, LOOKING NORTH.

Russian Valley. An hour's walk led to the crest of a rounded ridge, the northward prolongation of Mount Starashchin, whence we looked down on a lake, frozen over and snow-covered, whose mottled surface looked like a bed of morning mist filling a valley.

In the evening our men returned to the ship with some birds and the skin and blubber of a young whale. Later, a small boat came alongside, containing a Norwegian, Klaus Thue of Tromsö, and a Lapp, the two survivors out of a party of four that were caught by the ice in the preceding October, and compelled to stop in Spitsbergen for the winter. They had a horrible tale to tell of privation, sickness, and death. Thue himself was just recovering from scurvy, and looked very ill and weak. He related his story with the assistance of a detailed journal, which, I believe, was afterwards published in the *Aftenposten* of Christiania. It was briefly as follows :—

"Last year we came up in a cutter from Tromsö to shoot reindeer. There were four of us, the skipper Andreas Holm, Anton Neilsen, the Lapp, and I. We filled up with about seventy reindeer, and on the 14th of October we sailed from Advent Bay for Bell Sound, whence after a few more days' hunting we meant to sail home. On the 16th we were off the mouth of Bell Sound, and were surprised to find it blocked with ice so early in the winter. Moreover there was the ice-pack outside to the south and west, so that it was only off the mouth of Ice Fjord that there was open water. Then we knew that we were shut in, and must spend the winter in Spitsbergen. We sailed back to Ice Fjord, and on the 19th cast anchor in Advent Bay, just where the cutter now lies frozen up. The ice has since crushed her, and she is a wreck. She was an old boat and not strong.

"We lived on board till November 11. It was then getting very dark. On the 12th we shot a reindeer. That was the last day there was light enough for shooting. As soon as we reached Advent Bay we set about making preparations for the winter. We dug an oblong hole, about four feet deep, in a dry place a little above the hut. We could not dig deeper, the ground was so hard. We made a roof out of spars and sails. We used the door and other materials from the cutter's cabin, and brought the little stove and other furniture from her and arranged them in the hut. At first we burned the coal from the boat which we had got in Green Harbour. That gave out on the 5th of January, and then we fetched more from the hill near the hut, but do what we might we could not keep warm. The cold was dreadful. With all our furs we were never once warm during the whole winter. The ground beneath our feet used to bang like a cannon, cracking with the frost. Later on we made our hut better and stronger, but at this time it was a very poor shelter. Of course there were quantities of snow on the hills and all about, but there was never any snow on the flat of Advent Point—it was always blown away from there.

"Because of the cold we thought it would be better if we could move across to Nordenskjöld's House at Cape Thordsen, on the far side of Ice Fjord; so on the 11th of January, when the fjord was frozen over, we all set forth to cross the ice on foot. It took us fourteen hours to walk across. Anton Neilsen was already ill with scurvy. On the way across, his nose, eyes, hands, and feet became badly frost-bitten.

"We stayed in Nordenskjöld's House, which was much warmer than our hut, about a fortnight—till January 26. Then we had to go back for more provisions. We left Neilsen at Cape Thordsen with food enough. Bad weather came on after we left, and for twelve days we could not return; then, however, it cleared and we went back, dragging a sledge laden with food. We found Neilsen in a very bad way. His nose had dropped off; his feet were black, his hands almost black. He was also very ill with scurvy. On February 20, we had again to go for provisions. We left him with three weeks' food, but he seemed to us to be nearly dying, though we expected to return before he died.

"About the middle of the fjord we came upon open water and saw some walruses. Because of the open water

THE TOMB OF THE SKIPPER.

we could not go straight on, but had to follow beside a lead, which two bears were also following. We all landed near Hyperite Hat, and there we killed the bears and left them, meaning to go back at once with our sledges from Advent Bay, and take the bear's meat over to Cape Thordsen, for we had made in all three sledges out of boards and barrel-irons; two of these were at Advent Bay, and one was at Cape Thordsen.

"Bad weather again prevented us from crossing, or even from going for the bears. The ice broke up in the fjord and it was all open water, much more open than it is now. In fact from February to May there was no ice in Ice Fjord, but only open water. Our coal now gave out again, but we fetched a supply of three and a half barrels full from the little valley behind Advent Point. The bad weather continued till March 13, when we went in our open boat for the skins of the bears. The big one was still quite good, but the smaller had been torn by other bears. We tried to go over to Cape Thordsen, but loose ice came packing down on us, and we had to return to Advent Bay.

"Now the skipper became very ill with scurvy. He hobbled about on crutches for a time, but at last could no longer walk. For twenty-seven days he lay in his bunk and then died. The ground was so hard we could not bury him, so we put his body into two molasses barrels and covered them with a sail. He died on the 30th of April. On the 11th of May, we left Advent Bay in our open boat to go to Bell Sound, for I was stricken with scurvy, and we thought that some one would come into Bell Sound to look for us, and so we should be sooner relieved than if we stayed in Advent Bay. There was no ice anywhere about. We reached the cape at the north of Bell Sound safely, but were kept there by bad weather for five days. Then we went on to the Mittel Hook, where we set up a tent. I lay in it very ill for a month, and thought I should die. One day the Lapp saw a seal, and he ran, carrying a pail with him, and shot it and caught the blood in the pail. I drank that and immediately began to revive. I shall now get well.

"We did not trouble to go again to Cape Thordsen, because of course Neilsen has long been dead. I kept a journal from day to day, but some of it is in Nordenskjöld's House, where no doubt is also the body of the dead man. When we reach Advent Bay we will show you our hut, and then you will see that everything is as we have told you."

I wrote the tale down from Thue's lips, but have necessarily curtailed it, for he made many digressions, and was constantly referring to his journal or appealing to the Lapp for confirmation of the truth of his statements. It only remains to add that later in the season Mr. With went over to Cape Thordsen, but found no traces of the dead man. Perhaps after exhausting his food he wandered forth to try and cross the bay, and fell through the ice, or perhaps he died in the open air and was eaten by bears. No one will ever know.

The same night that these men were taken on board, the steamer *Virgo* anchored alongside of us with Herr Andrée and his balloon on board. She took on Mr. Stadling and such of our passengers as were going to join the balloon party, whilst we took from the *Virgo* the party of Swedes —Baron de Geer, Lieutenant O. Knorring, and their eight seamen—who were coming to make an accurate survey of the shores of Ice Fjord, the existing chart of which is so faulty. The *Virgo* presently sailed for Danes Island. She had left Tromsö two days before us, but was delayed by heavy ice.

The morning of the 20th was cloudy and grey; ice conditions, however, were favourable to advance, so we steamed ahead at an early hour. Navigating slowly and with great care, the mouth of Advent Bay was steadily approached. Far inland we could not see, but enough of the southern shore was visible to reveal the character of the country in that direction. It consists of soft, horizontally bedded rock, divided roughly into two stages or cut-down plateaus. Of the upper plateau only fragments remain, forming the higher mountains, such as Mounts Lindstrom and Nordenskjöld. The lower plateau, which was probably at sea-level when the upper plateau was being cut down, is represented by more considerable remains, but it also is divided by numerous valleys, and largely cut away. There remains no continuous plateau over which you can travel, only a series of broad ridges all of approximately the same height.

Winding her way amongst floating ice-blocks, the *Raftsund* ultimately reached the mouth of Advent Bay, which is rendered narrow by a low spit of land that juts out from the west and makes the harbour so safe, for ice cannot drive directly in upon a ship that lies behind the spit. The steamer entered without difficulty, and came to anchor between the fast ice of the bay and Advent Point. Our first goal was attained. We landed at once with Thue to visit the winterers' hut. Their poor little vessel lay near the shore, firmly fixed in the ice and full of water—doomed to sink as soon as the ice broke up about her. On the beach were such articles of value as they had saved—chains, anchor, reindeer carcasses, antlers, and so forth. The reindeer were in a frozen heap. There were also five barrels of reindeer meat, to which bears had recently helped themselves, for their great spoor was plainly visible in the snow. On a raised beach, some thirty feet higher, was the hut, and by it, in two barrels, the body of the skipper, close to some old graves. The hut was neatly built, dug out of the ground about a yard deep, and with trim steps leading down to it.

The end was formed of wood admirably joined, and with a door that shut closely and locked. There were some small glass windows, and a fireplace with an iron-pipe chimney. These things were brought from the ship. Within all was perfectly tidy—a bed along one side, various utensils, all clean and proper. The seamen's chests were ranged on one side, and on each was a label praying the finder, "for pity's sake," to send it to the owner's wife at such a place.

All the camp sites we could see were equally exposed, so we decided to pitch our tents by the Swedes, close to the edge of the fjord on a sandy and gravelly flat, dry enough, but not offering very good holding for tent-pegs. After dining on board, we began landing our goods, and set up the tents about a little square. There were six in all. Pre-eminent for size was Studley's African marquee. The others were three Whymper tents with flies, and two smaller ones intended mainly for use inland. Each of these was seven feet long by five feet wide and less than five feet high. They were of the Mummery pattern, made of strong Willesden canvas, rather heavy, but perfectly waterproof. For supports we used two ice-axes with pieces of wood made to fit on their ends and lengthen them to proper dimensions. The weak point about these tents is the number of cords required to hold them out, and consequently of pegs to hold the cords. In a rocky country this does not matter, because stones can be used, but in bogs one must have pegs. Three men in one of these little tents made rather a tight pack, but there was plenty of room for two and their small baggage.

The view from Advent Point camp was not seen to best advantage in the grey light beneath low clouds, though across the fjord to the large glaciers of Cape Boheman it was now and always lovely, and the floating masses of ice, that drifted

THE SURVIVORS AND THEIR HUT.

in and out with the tide close before us, made a novel and beautiful foreground. But our eyes most frequently turned in the opposite direction, for there lay our fate, our route inland. All we could see was the frozen surface of the bay and the white banks on either hand, with a valley bending away beyond, to the east, between white hill-slopes under a roof of cloud. I was destined never to behold this view clear. The hills were always buried in cloud when I was at Advent Point.

It was delightful to be again under canvas, living in pure air with no dark roof to shut out the sky. I sat till midnight writing in the door of my tent, and needing no extra clothing. The air was fresh indeed, but its freshness served only to make the blood course warmly through the veins. Late at night a shot was fired from the steamer—at some bird, I suppose. Gregory, half asleep, leaped up. He thought it was the Masai coming to loot his camp. How varied are the excitements of travel in different places!

STRIKING CAMP.

CHAPTER V

ADVENT BAY TO CAIRN CAMP

EXPLORATION brings keenest delights in its train; it has also detestable concomitants, most of which may be summed up in the word, baggage. The problem of exploration is to get the baggage over the ground; it is easy enough to transport one's self. Thus the importance of equipment becomes manifest. Its virtue is to be as light and compact as possible, consistent with the functions it has to fulfil. Our equipment was in reality compact, yet in actual amount it was large. Most of it was to be stored at the base, for we could only carry two sledge-loads inland at any one time. Now every individual object had to be landed, piled, combined, sorted, distributed. Nearly a ton of hay, oats, and beans had to be carted up from the shore and stacked. Food, sugar, salt, knives and spoons, and every little detail had to be unpacked, divided, and made ready for use at any required moment. All this was the work of hours many and long. Moreover the *Raftsund* would presently sail away, taking our last mail for some weeks; letters therefore had to be written home, bills on board paid, and all preparations completed for being marooned as com-

ADVENT POINT CAMP.

fortably as possible. The work was done in the open air, whilst cold showers fell, and gusts of wind blew shavings, dust, and every light thing hither and thither in odious errancy.

The worst trouble was to land the ponies. A small barge had been brought up for this purpose. It was hoisted off the deck by the donkey engine and swung overboard. For a moment it hung in the air, then, with a loud report, the chain snapped and the barge fell and broke asunder in the midst when it dashed upon the water. So a raft had to be made out of planks laid across two whale-boats, and on this platform the trembling animals were safely brought ashore one by one. Great was their joy to feel solid earth beneath their feet. They kicked about with glee and rolled on the ground, then sniffed for grazing, but found little satisfaction on the stony flat, which in a few weeks' time was to yield them so succulent a harvest.

The crew of the *Raftsund* worked like fury to get rid of the materials for the hut, for the drift ice greatly increased in amount, setting across from the north and packing into the bay in a continuous stream. The steamer's position had often to be changed, and men in boats poled the big masses away from the neighbourhood of the screw and kept the ship free. At one time it almost looked as though she would be beset, but the tide turned and the danger passed. It was a real danger. A week or so later, when I was inland, I am informed the ice again set in much more heavily and almost filled the bay, so that, if the *Raftsund* had then been there, she would assuredly have suffered. Overboard went the hut's materials, the planks into boats or on to the raft; the beams point downwards into the water, plunging right below the surface with a great splash, and almost leaping out again with the recoil.

We named the ponies Spits and Bergen. Spits was a

mare of very phlegmatic disposition who soon made herself quite at home. Bergen was a nervous beast, and signified his dissatisfaction with his new surroundings by presently breaking away and careering over the flat in wild terror. When recaptured, he was trembling and foaming with fright—a bad beginning. I noticed, lying on the beach, a small sledge, the property of the winterers, which Klaus Thue had made out of boards and barrel-irons. He gladly sold it to me for a few kroner, and it served us well, being far better suited for inland work than the costly machines we brought with us from London. Had we then known what we soon afterwards learned, we should have sent back to Tromsö for a few more of the same stout pattern. Garwood at once harnessed Spits to this sledge and drew our goods on it to camp. She settled down to the work and did it well.

Our preparations were watched with a somewhat embarrassing interest by the steamer's passengers, who landed one by one and took snap-shots at us with their cameras, not always immortalising our most dignified moments. One passenger, who came with us all the way from Trondhjem in order to see Herr Andrée, and had never once quitted the ship, now landed and gave infinite delight to every one. His costume was most picturesque—long boots, a long ulster, a great fur cap, a revolver slung round his waist, a horn over one shoulder, and a camera over the other. The horn, he explained, would be valuable if he were to be lost on the mountains, whose gentlest sloping foot he never approached. He walked up and down on the beach with dramatic gait, then turned towards the bay and solemnly fired off all the chambers of his revolver, after which he blew a blast on the horn. Then he fired off his camera in all directions, and so returned to the ship and disappeared.

Later on in the season, I am told, though I did not see

them, many strange tourists came and disported themselves strangely at Advent Bay, during the few hours that the weekly tourist boat used to wait there. They always brought rifles with them, under the impression that bears, or at least reindeer, herded at every point along the shore. There being nothing to shoot, they nevertheless fired off the rounds of ammunition in their little store, aiming at birds, or merely into the air. Many were the narrow escapes of inoffensive onlookers. A bullet came close over the tent of one of my companions. Others whizzed near the heads of the salvage men working at the winterers' wreck. One foolish creature is said to have mistaken a photographer with his head under a dark cloth for a reindeer, and put a bullet through his hat. Another, when we were away in the little steamer on the north coast, stalked, and I believe fired upon our inoffensive ponies. At last, for mere safety's sake, two little targets were set up between the inn and the sea. "What are these?" I asked on returning to Advent Bay. "They are the game the tourists fire at," was the answer. Poor things! they had been told that Spitsbergen was a sportsman's El Dorado. Asking for walruses and bears, they were given—targets!

Within little more than twenty-four hours from arrival, the *Raftsund* blew her whistle and began to move away, still pitching the inn overboard as she went. We watched her steering a very devious course through the ice, heading not for the sea, but due north across the fjord, where, after two hours of great anxiety, her ice-master found open water. Baron de Geer is my authority for the statement that such profusion of ice at this time of year is unusual. It was the more remarkable because, as we presently learned, the sea to the north and west of Spitsbergen was opener this year than it had been at any time during the memory of man. This ice, however, did not come from the north or west, but out of Wybe Jans Water, round the South Cape, where we saw so much drift ice, and so up the west coast. When Baron de Geer was in Spitsbergen before, at the same time of year, there was no ice in this fjord, and the snow was melted off the hills to a considerable height. Now snow was lying thickly down to the very water's edge, and two-thirds of Advent Bay was frozen up, as indeed were all the other arms of Ice Fjord.

All this day (June 21) clouds hung low above us, but upon Advent Vale, the valley at the head of the bay, the sun continuously shone, leading us to the mistaken belief that the climate of the interior might be better than that of the coast. Attracted by the brightness, Trevor-Battye and Studley started for a walk inland, up the west bank of the bay and the river at its head, nearly as far as where we afterwards pitched, Bolter Camp. They suffered many of the discomforts we afterwards experienced, and found no reindeer, no new birds, nothing that in any way pleased them. Studley described the country as one botched in the making and chucked aside unfinished. For his part, he said, he was "dead off it," and would get away the first chance he had. Trevor-Battye was not much less displeased. He came to study the birds of the country, and hoped to find much new matter of interest away from the coast, but he found nothing, and saw little promise of finding anything novel. Most of the birds of Spitsbergen haunt the coast, and there build their nests.

Meanwhile Garwood and I worked away at the baggage, helped by Ted in the intervals of sketching. We sneezed and shivered and pitied ourselves between whiles, but the work progressed. Gregory suffered worst from the chill, and became visibly unwell, so that he had to take to his sleeping-bag and feed on slops. This was the death struggle of a colony of African malarial germs within him. They were presently destroyed, and he became as strong as any

ICE IN ADVENT BAY.

one. The daylight night seemed yet more miserable, not that the temperature was specially low, measured by the thermometer. There is really no relation between thermometric scales and the sensation of cold. A blustering wind, edged with the keenness of the neighbouring ice, howled over the dreary waste. When the time for rising came, the warm sleeping-bags were hard to quit, but quitted they were, and work went forward, Trevor-Battye skinning birds, Studley excavating a huge fireplace, Ted sketching, Garwood and I disentangling the stores and supplies needed for loading the sledges. Sledge-loading was new to both of us and took an unconscionable time. By two P.M. one sledge was finished, the canvas cover laced over it, a pony caught and harnessed, and we were ready for an experimental march, exactly forty-eight hours from the moment when the *Raftsund* first cast anchor in Advent Bay.

We started along the shore, which here was terraced by a series of sharply-defined parallel ribs, following the curvature of the coast and very neatly sloped. They were made by the edge of the fast ice pressing against the beach and leaving its mark at different tide-levels. A belt of fast ice still hung by the shore, some hundred yards wide, all round the open part of the bay. We toiled over the flat ground, sinking deeply in from the first step, but rejoicing to observe how well the sledge slipped along on its ski runners. It was but a brief moment of satisfaction, for very soon came the wide bed of a torrent, or rather a whole series of channels made by the changeful stream, a ridged and furrowed stretch of large broken stones. Here the sledge began to bump and bang, pitching down into the channels and being hauled up the other side, the sharp edges of the rocks scraping and shaving the ash runners, and the unevenness of the ground straining the sledge in every direction. Clearly, if there was to be much of this kind of work, the sledges would not stand it. But we saw long stretches of snow ahead, and believed that they would form the chief part of our way.

Thus the north-west corner of the bay was gained, where Advent Point springs from the mainland. Now we turned south, ascending to traverse above a long low rock-precipice that borders this part of the bay. The traverse was over snow-beds and bogs, into which Spits sank up to the belly. Garwood led or rather dragged her along, for we had no whip or other stimulant. Pedersen and I hung on to the sledge to keep it from sliding sideways down hill. All went merrily enough till we reached our first gully. The slopes of the soft crumbling hills in this part of Spitsbergen are seamed with gullies cut out by streams that carry off the melted snow from above. As yet we had no conception of the amount of melting accomplished in the long spring and summer days in the Arctic regions. When the thaw sets in, and lasts all twenty-four hours of one day after another, it reduces the snow for a considerable depth to mere slush. The ground beneath remains hard frozen and slippery till the last fragment of snow upon it is gone, then it too thaws and turns to bog. The amount of water that pours off the hills and floods the valleys, during the month when the melting is at its maximum, is enormous, and denudation is correspondingly energetic. Thus a stream, which by the end of August will be dry, digs into a slope a gully of astonishing depth, or forms on the flat a fan of wide extent, rendered irregular by countless channels. It is unnecessary here to enlarge upon this statement, for the course of our journey made us intimately acquainted with all the effects of denudation upon an unexpectedly large scale.

The hill-side gullies at this time were filled with snow to the brim, and the streams flowed down tunnels beneath. In fact, the gullies were snow couloirs of gentle slope wherein the snow as yet retained a certain consistency. At

PONIES STUCK IN BOGS AND SNOW.

the first we unharnessed the pony and led her over, then ourselves dragged and lifted the sledge. The next gully was treated with less respect, and Spits did the hauling with so much good-will and kept her footing so skilfully that we relied upon her cleverness to negotiate all the snow-couloirs that succeeded. Three valleys leading westward opened one after another. In the drainage débris of each were fragments of coal. The swamps about their mouths were laborious to cross. Fog filled all these valleys, so that we could not see up them; later on we learned that they are merely short trenches eaten back into the old plateau by vigorous streams, some going farther in than others but none being true orographical depressions. Once only this day we just discerned the edge of the snow-field on the plateau, curling over the top of the slope in a huge cornice.

Three hours of such work covered about five miles, and was all the pony could accomplish, so we hauled the sledge on to a dry knoll, and set forth to return to camp a good deal wiser and somewhat less enthusiastic than we started. Now that we could have eyes for the view, there was little to be seen. The bay, almost wholly covered with ice, spread abroad beneath the low cliff under a roof of grey cloud, which obscured the hills and saddened the valleys. Everywhere the eye rested on barrenness and desolation. The sighing of wind and the cries of geese by the water's edge alone disturbed the silence of this abandoned place. We returned in our steps, warmed and enlivened by the work done, and presently cheered by a burst of sunshine. It came over us by what seemed to be the grave of a sailor, a mound framed in a ring of stones bearing a board thus inscribed—

KAPT. VOGELGESANG

S.S. *Columbia*

HAMBURG.

D. 29 . 7 . 1893

We wasted much valuable sentiment on this supposed unfortunate, and only some weeks later learned that the monument was raised to commemorate a gigantic beer-drinking bout or *Kneipe* enjoyed by the tourists who came up in the great Hamburg-American liner on the date mentioned. Other Kneipe-monuments were erected beside this one during the course of the summer—gaily painted and inscribed beer-barrels, iron flags, and so forth. Assuredly the vulgarisation of Spitsbergen has begun.

The view from this spot is indeed the best in the immediate neighbourhood of Advent Point. When the glaciers of Cape Boheman are clear, away across Ice Fjord, and the water lies blue beneath them, it must be superb. It was charming even on this dull day, when fog and cloud enveloped the distance, for the outline of the bay was graceful, and the margin of the smooth ice curved harmoniously from it. Sunlight carpeted the mottled frozen floor, except where a cloud cast a blue shadow down into the midst, whilst the floating ice-blocks glittered off Advent Point. On the plateau, where the ill-fated crew wintered, the Norwegian flag waved over the hut, in honour of the dead man whose body had just been consigned to its last resting-place in presence of the Swedes and some of the members of our expedition. When camp was reached the cold wind blew again and the sun was hidden. Larger ice-floes than ever came drifting past the point. Our last night by the shore was depressing enough.

Next day (June 23), we were to make our final start inland. It was neither fairly dry nor tolerably fine. The loading of the second sledge was at last finished. It was burdened down at one end with half a truss of compressed hay, whilst various objects, remembered almost too late, were tied on outside the canvas cover in picturesque confusion. Garwood and Pedersen were to drive the sledge; I was to set forward alone with the plane-table and begin the sketch survey. Gregory walked a little way with me, full of regrets that he could not at once come farther, but he was not yet cured of the effects of his chill. He sat on Vogelgesang's beer-trophy, while I set up the instrument for the first time.

The theory of plane-tabling is so simple, the instrument so devoid of complication, whilst in practice at home the working seems so easy, that the actual difficulties encountered when a new country comes to be surveyed are not readily imagined by one who has no actual experience. "I suppose," said Trevor-Battye to me, "the plane-table work is very easy." As a matter of fact it is often very difficult, but nowhere more so than in Spitsbergen, where it is heart-breaking. There, in the broad valleys and featureless slopes, it is practically impossible to decide from one station what shall be the position of the next, or from that to identify the preceding. The compass moreover does not enable you to orient the table properly, for the hills are full of iron-ore, which deflects the needle in the most changeful manner. If one could carry a theodolite, and take occasional true-bearings, this source of error would be removed, but for this observation the sun must be visible. In Spitsbergen one seldom sees the sun. Again, the hill-tops are for days together covered with clouds, so that it is the exception to gain a second sight of the whole series of points observed from previous stations. Frequent showers wet the paper and wrinkle its even surface, and then it will not dry for hours, so that you cannot ink in your sketch while details are fresh in your memory. Thus estimates have often to take the place of observations, unless you can afford to wait upon the weather, as an explorer never can. The inaccuracies introduced one day can seldom be corrected on another, and thus perplexities multiply. The quality of the work finally produced depends upon the alertness of the traveller, upon his keeping his eyes constantly on the watch for momentary glimpses, which may reveal the structure of the country and by degrees build up in his mind a clear conception of the forms and relations of mountains, ridges, and valleys, to which such accurate observations as he may have been able to make serve to give precision. When the journey is ended, and the map redrawn at home, it often happens that the particular photographs which were relied upon for certain details prove to be failures. All your other blunders and omissions then become apparent, and ultimately what you get for your pains is a survey in which any fool can detect errors and manifest them as proofs of your incapacity. Yet surveying a new land, with all its troubles, possesses great fascinations. It is delightful to behold the blank paper slowly covered with the semblance, however vague, of a portion of the earth's surface before unmapped. The interest of every view is increased when it has to be analysed structurally. Each mile traversed explains the mile that went before. Each corner turned reveals a tantalising secret. Every march solves a problem and leaves in the heart of the surveyor a delightful sense of something accomplished.

Gregory and I parted as we saw Garwood approaching with the tandem of ponies dragging the loud-complaining sledge. Both animals worked well, the timid Bergen on the whole better than phlegmatic Spits, who sometimes jibbed. Gaily the two of them hauled the sledge over the shelves of the terraced slopes above the west bank of Advent Bay, ploughing it through the snow-couloirs, and tearing it over slopes of rock-débris, which scraped away the edges of the ash ski-runners far too quickly. When the second sledge was reached, the loads were rearranged and on we went.

All day long the light was pale and feeble, like that of a cloudy English afternoon in December. Cold showers fell

ADVENT BAY IN AUGUST, BUNTING BLUFF IN THE DISTANCE.

with increasing frequency and in increasing volume. For two hours we toiled across slopes, so boggy that the ponies sank into them up to their bellies, and once both stuck fast at the same time, seated on their haunches long enough for Garwood to take their photographs. Beyond the cliff a descent was made on to the flat ground near the head of the bay. It was thought that the worst of our troubles were now over, and that henceforward the route would be easy as far ahead as could be seen. How little we knew!

The exact position of the head of the bay was not discoverable, for the river empties into it through a hybrid region of land and water, land that at the best is swamp, and water that is mud. Now land and water were alike enveloped in snow, between which and the frozen surface of the bay was no line of demarcation. We kept beyond the snow on a portion of the great bog which fills the whole valley bottom hereabouts. The summer thaw had as yet only penetrated about a foot deep into the spongy ground, sometimes less, and there were patches covered with ice, good to travel over, for the ponies' shoes were fitted with long spikes. But the frozen islands ended all too soon, and the snow beds were utterly soft and had to be waded. Many side-valleys opened, all leading up to the plateau between Advent and Coles Bays. At the mouth of each was a many-channelled fan of stone-débris, about half a mile wide. All the channels were filled with snow-slush or running water. They had to be waded, one after another, with the water washing right over the sledges, the contents of which were protected by Willesden canvas covers. On these occasions the ponies kicked and floundered about. Their work was very severe, and they were as yet in poor condition.

We soon became callous to wet, fortunately before reaching the worst place of all, where there was more water than land, and the moss merely stood up in islands and ridges out of the water. Rain poured persistently, and all the hill-tops were lost in cloud. Evil conditions for surveying! One long snow-bed, through which a stream flowed, was powdered over with broken coal. Bits of coal lay about everywhere, and the bogs were in places black with it. Once I stumbled over what I thought to be a stump of wood. Looking back to see how such an unwonted substance could be there, I found it was a reindeer's antler. The head was so buried in moss that only the tip of a tine emerged. We found plenty more in the same condition, for the whole interior of Spitsbergen is strewn with antlers. Shortly afterwards reindeer footprints appeared, and we presently sighted two deer, looking grey in the distance against the black stones. It is generally said that, in Spitsbergen, reindeer are so tame that they walk up to you to be shot. This is by no means always the case. These deer when they winded us went off up a side valley, and Garwood could not come near them in the time at our disposal. Plenty of bernacle geese flew about or honked at us from the swamps. Often they sat still and let us come within fifty yards of them. But we had no time to attend to their challenge; we only thought of forging ahead.

At ten P.M. the mouth of the fourth side-valley was reached. Here it was necessary to pitch camp (Bolter Camp, 230 feet). A site was chosen on a bit of rising ground at the edge of the swamp. It was spongy and damp as could be. Resting the elbow on the rubber sheet, as one lay on the ground, it sank in and made a hole. But there was no drier spot within range, whilst here was grazing for the ponies—a matter of importance considering the small amount of fodder we could carry, for it was already apparent that our ton of hay and oats at Advent Point would be mostly wasted.

The little tent was soon set up with a sledge on each

side to tie it to instead of pegs. The ponies were hobbled and let go. We took off our wet clothes and crawled into the reindeer-skin bags, which are as gymnastically hard to enter as they are morally hard to quit. The spirit-stove was lit, ration cartridges cooked, and tea made. We supped with deep satisfaction, knowing that we had grappled with the real work of our journey, and that the first and, we flattered ourselves—how erroneously!—the worst stage was accomplished. Supper and rest had been earned. By two A.M. journals were written, and we could settle down for sleep. Side by side we lay, Garwood, Pedersen, and I, each in his bag, looking like three large dirty white cocoons. Heads were inside the bags, partly for warmth, more for darkness' sake. Thus protected, none ever suffered from cold; sometimes we were over-warm. Rain poured soothingly on to the roof, but the poor ponies suffered and came up to the tent seeking shelter. They kept stumbling over the guy ropes. Driven off again and again, they constantly returned.

The wind was from the north-east, a bad quarter, said Pedersen, who prophesied fine weather with a south wind. Experience taught us that all winds bring rain in Spitsbergen. If the weather means to be fine, it is fine whatever the wind; usually it is foul. Next morning (June 24) it was impossible to start, for no surveying could be done, everything being enveloped in fog. We had to stay where we were till the position of the camp could be fixed. So we sat up in our bags and were fully employed over details of all sorts—changing and packing photograph films, cooking, and so forth. In a tiny tent there is always much to be done, for everything takes an inordinate time to do. Whatever can upset does upset. Things become immediately mislaid, being necessarily piled one on another. Order cannot be maintained without a little space. Where three men are crowded together in the area they just cover, every search for an object involves gymnastic gyrations. I concluded to add a "lazy tongs" to Gregory's list of camp indispensables.

Venturing forth with bare feet to inspect the weather, I stumbled into a filthy bog-hole for the first of many times. Our wet clothes of yesterday were as wet as ever. Clouds came yet lower and rain never ceased. The day was passing and nothing was accomplished. We filled ourselves with ration cartridges, and pitched the plates out to be washed by the heavens. Instead of that Spits licked them clean, and then pushed her head in at the tent door and made a face at us, curling up her lips into what looked like a complicated sneer. At last there was a brief clearance. An unsatisfactory observation was taken and camp could be struck. The sledges were nearly loaded. Spits was just harnessed, Bergen being harnessed, when he gave us the slip and went careering around. There was no catching him. Picture our disgust! Little was said; the work of preparing the sledges went on. We hoped the beast would return to the oat-bag, but his memory of the flesh-pots of Advent Bay was a stronger attraction. In a few minutes he became small in the distance. Pedersen was accordingly sent off to bring him back. He was found at Advent Point, which he reached just in time to be harnessed to another sledge and made to draw coal from the neighbouring hill.

Our prospects were dismal. There was no redeeming feature. We determined to make a half march with the remaining pony and one sledge and then to return for the other. The weather remained abominable—rain falling from low clouds in the midst of the bleakest surroundings imaginable. How it came to pass that the interior of Spitsbergen had remained for us to explore was no longer a mystery. The job was more laborious than any of us anticipated. So much the better, had we been properly

CROSSING A FLOODED TORRENT.

equipped, but now we knew that the sledges were, for the purpose in hand, mere costly ineptitudes. They were breaking down at all points, the runners so scraped and torn away that it seemed probable they would vanish in two more marches. To contemplate our problems was mere waste of time. They had to be solved by action, so about four P.M. we set forth, descending from our camp promontory on to the wide stone fan at the mouth of Bolter Valley.

If the previous march was bad, this was infinitely worse, for the reason that the farther we went inland the more backward was the season, and the more frequent, deep, and utterly rotten and sodden was the snow. The worst of it was that we might better have stayed where we were. Our present plan was to form a camp at some suitable point inland, whence we might make expeditions in various directions, and more especially one over to Low Sound. Pedersen, repeating the traditions of reindeer hunters, said that Bolter Valley led to Coles Bay, and that the next (Fox Valley) was the first that led towards Low Sound. In this, as in everything else he told us, he was inaccurate, the fact being that the reindeer hunters know little about the interior beyond a very few miles from the coast. Bolter Valley divides a short way up, as, but for the fog, we should have seen. One branch goes almost due south to Bolter Pass, and so to the valley of the Shallow River and Low Sound; the other leads west over the plateau to the Coles Bay Valley. From Bolter Camp we might have explored both routes. As it was, we did the best thing we knew of, and went pegging ahead up the left bank of Advent Vale, which here makes a great bend to the east.

A few minutes' trudge brought us to the edge of the many-channelled Bolter River, now in full flood with the melting of the snows in the large basins it drains. A man going alone might have picked his way across the channels and not gone in much above the knee. But one of us had to guide the pony, whilst the other held the sledge, giving it a twist now and then to avoid big rocks, or prevent it from being rolled over by the torrent. As a matter of fact, it did overturn in the deepest place, so we were both soaked to the waist in the icy water—a fortunate preliminary that made us callous to future wettings. Bogs of all sorts lay in our way, and infinite streams. Gregory, a few days later, counted the number in a single batch over a similar fan, not a mile broad, and found fifty-two which had to be waded, besides others small enough to be jumped. The ordinary bogs were of little account; one sank in up to the calf of the leg and floundered on. It was the snow-bogs that we grew to hate—deep beds of snow, smooth and solid-looking, but water-logged, and affording no particle of support. The foot went right through them to the solid ground, making a green hole that instantly filled with water to the brim. Whilst, however, it was impossible to stand on this stuff, great exertion was required to push through it. The pony sometimes sank into these snow-bogs up to her neck. She could only advance by a series of leaps. Sometimes she would fall over on her side and become stuck, sometimes she was held tight, seated on her haunches, with her forelegs beating the air. More than once we feared she would break her back, but she won our admiration by the pluck and intelligence she manifested, and by her consequently rapid development of skill.

All the march we were continually crossing below the mouths of little valleys, each discharging a stream over a many-channelled fan. Thus we rang the changes on streams to be waded, stone-slopes to be stumbled over, and snow-bogs to be floundered through. Each stream had snow-bogs for banks, making the scramble out of it a

ADVENT BAY FROM CAIRN CAMP.

matter of difficulty. At last came the very large fan at the mouth of Fox Valley, up which we proposed to make our farther advance. We had therefore to camp on a boggy knoll at its mouth, choosing this exposed position because the top was dry and visible from afar. We built a stone-man for further emphasis, and called the place Cairn Camp (340 feet).

Throughout the march there had been no view save of barren wastes and the feet of frozen hills, cut off at a height of about 500 feet by a flat grey cloud-blanket. Now, looking back down Advent Bay, we could still distinguish the Point, and far beyond it across Ice Fjord the glaciers and ranges behind Cape Boheman, bright under sunlight. That particular piece of country was far more often beheld by the sun than any other spot in the island that we saw. When it was raining everywhere else, sunlight gilded the Boheman glaciers.

It rained heavily when we began to pitch camp, and a fog came close about us. A white fox sneaked up to watch our doings, glaucous gulls flew by and faded away in the mist, purple sand-pipers uttered their plaintive cry, and the murmur of many waters filled the air. We fed Spits, and rested ourselves under shelter, with wet legs stuck forth beyond the mackintosh sheet. A short excursion up Fox Valley revealed difficulties, and induced doubts as to whether that was the right way to Low Sound. It was a bleak winding valley with a narrow stony floor, broken by a stream in a deep gully, whose precipitous walls were fringed all along both sides by great snow cornices. The only way ponies could take sledges up it was along the bed of the stream, where ours would be smashed up in a short time. Clearly we must climb to some high point and take a good view round, before a decision could be come to about what should next be done. The only food on the sledge was a valuable but unattractive substance called Bovril Emergency Food. It was so securely soldered up in a strong tin that, on an emergency, it could not have been opened. We smashed one side in, and cursed the man that packed it.

The return journey was dull and depressing to all but Spits. She rejoiced in the erroneous belief that she was going home and to Bergen's company. The clouds were no less dense than before. The mournful boggy slopes seemed sadder and more deserted. The character of the scenery is understood when it is perceived how the surfaces of all the hillsides, softened by frost and wet, slowly flow down, as bog glaciers, to lower levels. Thus all except a small portion of what once was Advent Bay is now a muddy plain, and this in turn is being overflowed by bog accumulations along the foot of the hills.

Bolter Camp was reached in less than three hours, Spits quickening her pace at the end, and hurrying up to the second sledge; she looked round and neighed, but no answer came, for Bergen was far away. After a brief halt and a meal of lime-juice nodule and biscuit, the second start was made. The way seemed longer, and Spits became worse bogged than ever in deep snow slush. She was too fatigued to extricate herself. Standing waist-deep in the green and freezing compound, we had to dig her out with our hands. The final pull up to Cairn Camp almost overtaxed her powers, but when the goal came in view she made bravely for it, and was rewarded on arrival to the best of our ability. Unpacking and cooking occupied another two hours, and then at last, about five A.M., we could turn in and enjoy the soothing patter of rain upon the tent roof, almost the sweetest music that can introduce slumber.

Printed by Lemercier Paris

MOUNTAINS BEHIND CAPE BOHEMAN FROM ADVENT POINT

CHAPTER VI

ACROSS THE MOUNTAINS TO LOW SOUND

BY this time our occupations no longer bore any relation to the indications of the clock. All twenty-four hours round there was the same effect of full daylight; the temperature between midnight and eight in the morning being perhaps lower than at other hours, but with no appreciable dimming of the sun's light. Thus we started for our marches, halted, cooked, and slept whenever was most convenient, the tendency being to lengthen out the days, as measured between intervals of sleep. The freedom from fear of benightment is a great relief. The corresponding lack of stimulus to hurry is a loss. On the whole, the loss overbalances the gain.

Our plan, for what must be called June 25, was to climb a hill anywhere to the south of camp, looking for a view, which the cloud-roof did not render a hopeless prospect, for it was a little higher than before and very much thinner, the sun even breaking through sometimes. Pedersen returned with Bergen from Advent Bay, and went off again up the main valley after reindeer, whereof two presently fell to his rifle. After laying out our things to dry, Garwood and I withdrew the ice-axes from their place as tent-poles, and, leaving the tent flat on the ground, set forth up Fox Valley, which opened to the south just behind camp. We followed its left bank over snow beds, bogs, and stony tracks, descending into and crossing two deep, precipitous-sided gullies, whose streams, covered by snow-tunnels, drain glacier-filled hollows in the mountain mass to the west. Beyond the

F

second gully we turned west up an easy snow-slope, then a more toilsome slope of débris, which narrowed to a ridge. Here Garwood discovered fossils in great number, and more were soon appropriated than could be carried, so we cast them on to the clean snow-slope below, where they could be found on the descent. While the fossils lasted progress was slow; eyes were glued upon the ground and the developing view was forgotten. Presently the débris became too steep and rotten to be farther ascended, so we turned aside to the neighbouring snow-ridge. I have not elsewhere seen débris slopes at such a steep angle as many we encountered in Spitsbergen. The reason is that the friable rocks there are smashed up so completely that the fine dust holds the fragments together, and, by increasing the friction between the pieces, prevents them from sliding. Débris slopes of such steepness are most laborious and even difficult to ascend; the surface is too steep to tread upon and not soft enough to yield readily to a kick. When such slopes are saturated with wet, the foot adheres to them as though glued down. On the snow-ridge were no fossils, so the view received attention. The weather was doubtless clearing. Advent Bay lay like a map beneath, with sunshine upon its burnished surface.

The snow arête gave place to a narrow cocks-comb of the rottenest rock imaginable. Fearing lest the whole thing should give way, we turned on to the right face, down which were falling frequent avalanches both of snow and rock. It became necessary, therefore, to return to the ruinous crest, which stood upon the ridge like a wall with vertical sides about forty feet high, so narrow that in several places it was pierced right through with large holes, and so rotten that huge masses of rock gave way at a touch, whilst at one point, above a hole, the entire mass groaned and trembled as we gingerly picked our way over. An hour was spent on this part of the climb. The final scramble was easy. On reaching what we had believed from below was the summit of a peak, we were surprised to find that it was merely a promontory between two gullies, jutting out from a wide, undulating area of snow, or rather of ice, with a snow blanket upon it. The ice-sheet, several square miles in area, sends Glacier tongues down many short valleys and gullies. It was the feet of these we had been crossing between Bolter

MOUNTAINS NEAR ADVENT VALE.

and Cairn Camps. Afterwards we made the entire tour of the mountain mass, by crossing Fox Pass, descending Plough Glacier, and returning by Bolter Pass. Standing here on the bluff, and looking across Advent Vale, it was easy to perceive how the plateau behind had once formed part of a wide-extending flat, which had been elevated and cut down into many valleys by streams eating their way back into it. We were, in fact, on what I have called the lower plateau, looking edgeways along it. In various directions we could see the remnants of the upper plateau standing upon it in the form of rounded snow-covered hills, such as Mounts Lindström and Nordenskjöld.

At the point of the bluff, when we arrived, a little snow-bunting greeted us in so cheery a fashion that we named the place Bunting Bluff (2480 feet) after him. The view in all directions was now brilliantly clear. Advent Point lay like a needle across the bay's mouth. A sloop off it was a tiny dot on the bright water. The large mountain area behind Cape Boheman, extending from Nord Fjord to Keerwyk, was visible in sparkling detail from side to side, with its three great glaciers flowing down between narrow mountain ranges. At the foot of one are large hills of moraine, visible from afar. The sunshine was warm. There was no wind. To live was joy. Our meal consisted of an Emergency tablet, a biscuit, and a lime-juice nodule! Burning a tobacco sacrifice, we returned thanks that we were not, as an Alpine party, burdened with the carriage of bottles of red wine and bulky foods. But how good a bottle of wine would have been, carried by some one else!

The plateau behind rose into a low wide mound that hid all the southern prospect, and would command a more complete panorama. We wandered up it, one after the other, rejoicing in the splendour of the sunlight on the snow, and leaving the rope behind. The full extent of our section of the plateau now became apparent. It bent away to right and left, sinking to Fox and Bolter Valleys, and ultimately narrowing to a mere ridge, at the far end of which, about two miles away, was a broad snow peak, shutting off the view we most desired to behold. The attraction was too strong. With mutual consent we set forth towards the peak. No one mentioned the rope, for who would expect hidden crevasses in so even an area. As a matter of fact, the place was a maze of crevasses, but the snow was so hard that we trod through it into very few. Plodding, plodding, plodding, the wide white area was at length crossed to the foot of the easy snow-ridge, where a brown patch emerged.

THE SUMMIT OF FOX PEAK.

We hastened to see what vegetation it might carry, but it was absolutely bare, and the rocks of it were weathered to mud. On its surface were many water-rolled pebbles, washed out from the substance in which they had been embedded. The single living inhabitant of the mound was a tiny flea-like insect, captured by Garwood. He enclosed it under his watch-glass, and it promptly took a ride round on the second hand, but before we reached camp again it was in minute fragments. The snow ridge stretched up ahead, faultlessly white in a newly-fallen mantle, and with its sloping line dividing the view *per bend*, argent and azure, but the blue of the sky was faint and delicate, wholly different from the strong dark tone observed from high elevations. Behind spread the wondrous snow-field, so gracefully undulating, and with the midnight sun shining brightly over it. Upon the surface of the snow lay a brilliant rainbow, caused by the ice-spiculæ which we had noticed to be peculiarly numerous and bright as we came along. This phenomenon was new to both of us, nor have I ever seen it recorded. It can only be seen when the sun is low and shines on névé covered with a powdering of tiny snow crystals. The ascent of the ridge was fortunately easy, for we had come without axes or rope, not intending a climb. There was a big cornice to avoid. We followed in the fresh track of a fox, who gave his name to the peak (3180 feet).

The view that burst upon us at the top was a revelation. All such views are revelations; it is the quality of their charm. We were on the watershed between Advent Bay and Bell Sound. Plough Glacier was at our feet,[1] joining into Dreary Valley, which led south to the larger valley of the Shallow River and Low Sound. A sea of cloud filled these depressions, spreading away for thirty miles or more

[1] I give to places the names they afterwards received.

to a remote range of mountains, golden in the low bright sunlight and clear soft air.

Eastward the view was limited by a series of rounded and uninteresting snowy domes. North-west were the splendid mountains and great glaciers of North Fjord rising above the bed of cloud that roofed Ice Fjord and Advent Bay. A great blue shadow plunged from our feet, and lay on the mist. At the edge of it was a small rainbow, haloing for each of us the shadow of his head. Sunwards the clouds shone like the sun himself; east and west they seemed softer and were less brilliant; southward they were dark and grey. The shade temperature was several degrees below freezing, but the sun was so warm that it made avalanches peel off the north face of the mountain.

Study of the view around us, and of that from Bunting Bluff, enabled us to perceive that the plans with which we started from London must be modified, for the nature of the country was utterly different from what coast-wise observers had imagined. We had planned to cross Spitsbergen from east to west along two or even three lines. We now saw that little information of value would be attainable by that method. The whole region within view was an intricate mountain country broken up by a maze of valleys and not containing any large level areas. To strike a mere line or two across it would be an ineffectual method of investigation. A closer study of some specimen area was required, and this could only be accomplished by making a series of expeditions to right and left, and by ascending to a number of points of view—a process involving much expenditure of time and a slower rate of progress.

With minds at ease, and a rich harvest of observation, we set forth to return. The rainbow on the snow-field was now before us. Sometimes the bending of the névé brought its apex within a few yards of our feet, so that the particular gems of ice could be identified that rendered up each tone of colour. It was as though we were in presence of some great goddess of old time, whose head was the radiant sun, and her necklace of countless rubies, emeralds, and sapphires. Her robe of snowy samite was sewn all over with diamonds, and her veil was the gossamer mist that lay along the valley. At three A.M. (26th), we were back at Bunting Bluff, and the cloud-bed was just below us. A midge buzzed about and found a final resting-place in a test-tube. The sun was so hot that we lay on the stones and dozed in the genial warmth, but hunger prompted a return to camp. Unwilling to descend by the rotten arête, we bore away west into fog, down a snow ridge where it broadened out to featureless slopes. White snow and white fog were the sole things visible. They melted into one another without dividing line. When the slope became rotten and steep, we bore away to the left and lost all knowledge of our position, save that the camp was somewhere to the north. The compass led down an easy gradient, and presently we found ourselves on a glacier, discharging down a steep gully—the second we crossed on the way up. Here at last was water. Long and deep were the draughts of it that I swallowed. Then down again, with here and there a short glissade, and so out on to the easy slopes where the old tracks were rejoined. Camp was reached at five A.M. Supper followed, and as usual took a terribly long time to prepare in our cramped quarters. By eight A.M. we were wooing sleep, and the cloud-roof was above our heads, dense as ever.

It was after two P.M. when we turned out. The day for me was to be a busy one in camp. The map materials required to be immediately worked up, there were journals to be written, and a letter for our comrades at Advent Point Camp, with a final list of things required for the inland journey and a definite plan formulated. Pedersen went off with pony and sledge to fetch the carcasses of his reindeer. Garwood had an important geological problem to work out on the slopes of Bunting Bluff. I remained alone in the chilly camp, blowing on my fingers and writing by turns, with only the undulating bog before me sloping up to snow patches and a low roof of cloud. The snow was melting with great rapidity at low levels. When the low snow is gone the bogs begin to dry, and the snow bogs at all events, which are the worst, utterly disappear. Thus conditions improve for inland travel, for which the month of June is probably too early. Such at all events was Pedersen's opinion, delivered to me when he returned with his reindeer, and related how he had fallen into a river from its snow bank, and was drenched through from chest downwards.

Many possible plans were discussed between Garwood and me. It was certain that Low Sound could not be reached by the ponies. Moreover, they must go down and take the sledges to be mended. If two days' wear and tear had so thinned the ski-runners, two more days would finish them. Back therefore they must be sent to have thick planks fastened under them. This would take three or four days, which Garwood and I might employ on a knapsack expedition, a thing freely undertakable when there is no possibility of being benighted, and one can sleep on the first dry patch when the sun shines.

After long deliberation a plan was evolved and a letter written to Gregory embodying it, and detailing the supplies required. We had discovered that it was a quicker matter, involving less expenditure of spirit, to broil reindeer in butter than to cook any of the concentrated rations. With plenty of spirit and butter one could therefore live well, and the balance of our stores required corresponding readjustment.

As I worked all day at these and other matters, I derived much diversion from the proceedings of a purple-sandpiper hen, whose nest was only a few yards away. It was hard to find, owing to its bare simplicity. She had availed herself of a crack in the moss, and had just dropped her four eggs point downwards into the hole; there was neither packing nor embellishment. When any one approached her place she would run away, looking almost like a rat, then flutter feebly along as though broken-winged. Having thus drawn off the intruder, away she went with a laughing chuckle.[1] Two skuas evidently had business near the camp, for they kept flying slowly round it, often swooping within a couple of yards of my head. I was also visited by three glaucous gulls and a ptarmigan, so that with the ponies to look after I was anything but lonely. About ten P.M. Garwood came in sight as a tiny point descending from the clouds down a distant snow slope. I immediately set hand to the cooking of a monumental repast of tea, mulligatawny soup, and grilled reindeer, with biscuits and jam to follow. The whole was ready when he came into camp, bulging and burdened down with fossils from the beds we traversed on the previous day. He had an interesting story to tell as we sat at supper in the door of the little tent.

In the afternoon of the 27th, Pedersen left with the ponies and reindeer carcasses for Advent Point. Garwood and I shouldered our knapsacks likewise, and set forth in the opposite direction. We carried an oilskin coat, change of stockings, an extra wrap each, a small tin of Emergency Food, thirty-four biscuits, some lime-juice nodules, four oz. of jam, two sticks of chocolate, a rope, matches, tobacco, pipes, snow-goggles, prismatic compass, aneroid, thermometer, and camera. Taking our ice-axes from the tent, and leaving it an amorphous mass on the ground, we turned

[1] We built a little stone man by the nest, and she soon grew accustomed to our visits, and would let us approach within a yard to watch her.

up Fox Valley, following at first our old tracks. A quarter of a mile had not been passed before a reindeer was overtaken, a noble fellow with a grand pair of horns, not in velvet. He could have been stalked easily, and as it was, we almost came within range of him. Continuing along the west bank, we gained the foot of the glacier at the head of the valley in about an hour of bog-walking and débris-stumbling.

This, the Fox Glacier, is fed by a considerable cirque of névé, which stretches back to the west as far as the arête by which we mounted Fox Peak. It divides into two branches, and there is a col at the head of each, one adjacent to Fox Peak, the other and lower more to the east, and visible from Cairn Camp at the apparent head of the valley. We chose the higher col as more direct, and named it Fox Pass. The route led up what seemed to be névé, but proper glacier ice presently emerged. We had not yet learnt that there are no true névés in Spitsbergen. A batch of ice-crevasses smothered in snow gave some trouble. We turned most of them by bearing up the slope of the peak between the two cols. It was a steep slope, but all the new snow had fallen in avalanches off the lower part, leaving a brown edge above, which looked like a bergschrund. The true bergschrund was, however, still higher up. Keeping round the slope we reached our cold and windy col (2550 feet) about a quarter to eight P.M. Fox Peak to the west was covered with cloud; the mound to the east seemed a mere heap of fine débris which the wind blew on to the snow, thereby forming dust-pyramids, in their turn snowed over. Thus a peculiar area of mounded and dirty snow was produced, the like of which we found in other places. The view ahead was meagre in the extreme. There was a glacier (Plough Glacier), trending down in a WSW. direction, and there were snow mountains beyond, vaguely discerned through changing clouds. Snow fell upon us, and the low fog thickened; there was no temptation to halt.

Going southward at first, we reached the trough of the glacier, then followed it down and through a mere white

DESCENDING PLOUGH GLACIER IN FOG.

chaos. Vague rocks sometimes dimly appeared. Now and again a curiously formed rock-splinter would be seen through a cloud-gap standing weirdly forth from a ridge. An infinite melancholy reigned. Fancy-engendered sounds sometimes struck the ear, as of a dog barking, a cow lowing, and the like, but falling stones and sighing winds were all that actually stirred the air. Presently we were below the thickest mist and saw further afield. The eye in its turn played tricks, figuring, for example, men carrying loads in casual lumps of rock, whilst the snow-outlined edge of an ancient moraine, high up on the opposite hill-side, looked like a mountain-canal to carry irrigation-water to Alpine meadows. The lower we descended the more incredibly bad did the snow become. We bore to the right off the glacier, but profited nothing on the snow-covered slopes. In we went, up to the knee—sometimes up to the waist. I tried crawling gingerly on all fours, a painful mode of progression. We came upon a surface that would just bear our weight, not stepping but creeping along it, each foot moved forwards a few inches at a time. For a yard or two there might be stones; the evil snow succeeded, less tenacious than ever. At length came a boggy slope with two very tame reindeer on it; it seemed a blissful highway, and led down to a flat stony bog, immediately below the ice-cliff with which this glacier ends. Garwood found much of interest to observe in the section of glacier displayed by this snout. It was advancing with some rapidity, not ploughing up the vegetation before it, but simply gliding on to it. The same was the case with other glacier snouts seen this day. All were advancing in the same fashion. Garwood climbed right up to the ice precipice in his eagerness, risking his life in the cause of science. I was glad when he returned and we could eat our Emergency Food at a safe distance.

The whole area intervening between this point and the place where the Plough Glacier's stream joins that from Bolter Pass, was covered with reindeer tracks, and we saw two more reindeer close at hand. They were little disturbed by our presence, and advanced within easy range to inspect us. For the remainder of the march we drove them like goats before us down the valley. After halting to take observations at the point of junction, and wading

TORRENT IN AN ICE-FOOT.

the stream from Bolter Pass, we came to what seemed a very curious region, just below the great bulging snout of a glacier, descending from the west, out of a large cirque cut back into the western plateau. Before describing the phenomenon which we here encountered for the first time, I may take this opportunity of mentioning that all the glaciers in Spitsbergen differ from Alpine glaciers in respect of their appearance of viscosity. Alpine glaciers look like flowing things when seen from suitable points of view, but Spitsbergen glaciers have a much more viscous appearance, they bulge over and spread out at the snout as if they were made of honey. They flow, too, down very gentle slopes, and their surfaces are always very flat; if curved at all from side to side, the curvature is too slight to be noticed. Their rate of flow may be much more rapid than that of Alpine glaciers.

This particular glacier was unusually steep, and therefore ended in a snout of unusual height, which bulged over in a threatening manner. At its foot was a curious icy area, unlike anything we had ever seen below the snout of a glacier in any part of the world. It was what we called, and what, I believe, is already known technically as, an ice-foot. An ice-foot is formed by the percolation of the glacier stream through the mass of snow that accumulates during the winter below the glacier's foot. This snow becomes completely sodden, and is frozen into a solid mass of ice. Often more ice is thus formed in a year than is melted. Later in the season we came across the most amorphous ice-masses, which proved to be remains of the spring ice-foot. After the ice-foot has been formed, the glacier stream flows over it, about the time of the early thaws, and cuts channels into it, which renewed snow-falls block, so that they are constantly being changed. Pools are thus made in places, and these at times freeze solid, generally in a roughly-crystalline prismatic form, the prisms standing, end on, close together and opening out as the summer proceeds, and the lateral restraint is removed by the thaw. At first the remains of these frozen pools are seen as domes of clear blue ice. Later, as the sides of the domes and the interstices of the prisms melt, the rods of ice separate from one another, and stand out like a sheaf of glittering crystals. Of course where a glacier is forming a terminal moraine, and is likewise advancing and retreating restlessly from year to year, the ice-foot is mixed up with moraine heaps, and a most chaotic distribution of ice, snow, water, and mud results. Sometimes the ice-foot becomes buried beneath moraine, and so kept from the summer's warmth, when there is formed what has been well named "fossil ice," the antiquity of which may be very great. Such fossil ice we met with in the Sassen, Esker, and Fulmar valleys. Here there was no such complication. At first we thought the ice-foot was a remnant of the glacier from some former stage of advance, upon which it was now pouring over in renewed volume; but the true character of the formation presently became apparent. The largest ice-foot we saw was below the Rabot Glacier in the Sassendal, whilst below the great Ivory Glacier towards Agardh Bay there was no ice-foot at all, because there was almost no river to form one, an ice-foot being a function not of the glacier but of the glacier stream.

So novel a phenomenon naturally interested us greatly. We wandered away from one another in pursuit of our investigations. It was an uncomfortable place to roam about in at this time of year, for there were so many water-logged snow bogs and streams full of snow, all with a white surface of deceptively solid appearance, that it was impossible to find even a fairly dry route across. Still Garwood managed pretty well, and arrived at the far side tolerably dry. I was less lucky or less skilful, though I always probed for footing with my axe, and generally found it not more than knee-deep in the snow-slush, which for wetness is no drier than mere water. At last I came to what was obviously a big snow-bog or slush-pool. Creeping gingerly to the edge, I bent forwards and probed its depth, expecting to find bottom not more than two feet down. But the axe went in and in, up to, then over its head. I could not withdraw it or myself. I perceived, in a flash of recognition, that I was bound to follow it. To avoid taking a header into the slush, in which it would perhaps have been difficult to bring oneself afterwards right way up—a truly ignominious fashion of drowning—I jumped in, managing to turn partly round and get one hand on the bank, thereby avoiding immersion above the neck, though the bottom was out of my depth. To climb out again was the work of a moment. Had I been less wet before this adventure it would have been more disagreeable. As it was, the incident passed with little comment. A brief interval of sunshine that followed seemed blissful by contrast.

Moraines and ice-foot passed, we plodded along the edge of the bog-slope that intervenes in all these valleys between the foot of the mountains and the gully or cañon of the river. The going, bad at all times, became worse every fifty yards or so, when a side-stream had to be crossed, for each side-stream ran down a gully of its own, filled with snow-bog that had to be tediously waded through. About one A.M. (28th) we halted during a fine interval to wring out pattis and stockings and empty the water from our boots. Garwood unsuccessfully tried to sleep. We were both dog tired, not with the length but with the toilsomeness of the way. Dreary Valley was very tame, with its long brown monotonous slopes on either hand, striped with snow-filled gullies, leading up eastward to a series of diminishing hills, and westward to the edge of a plateau. An enduring gloom of cloud and damp overhung and permeated the place. Puffs of cold wind chilled the marrow in our bones. Then rain began to fall, each drop like the touch of an icicle. I put on my wet stockings and boots, wound the wet pattis about them, and on we went again over slightly better ground, where was much grass and plenty of food for our ponies if only we could get them over the mountains. At last we approached the mouth of Dreary Valley, and found a grassy hollow with a clear little brook rippling down it in tiny channels. It was the most protected place we had thus far seen in Spitsbergen. Moreover there was a small patch of dry ground, big enough to lie on, the first square yard of dry earth we had come across since quitting Advent Bay. Here accordingly we determined to rest. We were utterly weary with the labour of the march, due to the quantity and condition of the snow. If, as we were informed, the season this year was a late one, we had that lateness to thank for the worst of our troubles. In August no doubt the pass might be crossed and the valley descended without trouble and in a very short time.

Our camping ground, though relatively sheltered, was not really a choice spot. It was a sloping trough about twenty feet deep and forty wide. There were snow-walls on either hand. Mountains protected us behind, but there was only a low ridge in front, over which we could see the prominent isolated mountain, which so attracted us as a probable point of view when we first saw it from near Bunting Bluff. In preparation for repose we changed our wet socks for the dry ones from our sacks, put on our extra wraps, tied the lappets of our Samoyede fur caps over our ears and round our necks, then lay upon the ground back to back with the oilskin coat about our legs, our feet in a kind of pocket

THE VALLEY OF THE SHALLOW RIVER.

it had at the top of its back, and its skirts tucked under us for a ground sheet. Sods dug from the ground with ice-axes formed our pillows. Water rippled past us within arm's length. We spread our frugal meal of Emergency tablets, biscuits, jam, and lime-juice nodules, and rejoiced at the unlooked-for comfort. It was not long to last. Cold puffs of wind soon found us out; so did two ivory gulls, which flew round and round close to our heads before settling on the snow-beds a few feet away, and discussing us long and minutely. Concluding that we were not yet ready for eating they presently left. Seldom have persistent efforts to woo sleep been less graciously rewarded by that fickle goddess than were ours during the time of our stay. We dozed sometimes, but never really slumbered. Then rain began to fall heavily. We endured it without a murmur; things had gone too far to be complained of, they were past the power of words. This I will say, that when your bones are aching and your position is so cramped that you cannot move, and the freezing ground is sucking away the little heat left in your frame, to be splashed in the face with icy rain is an aggravation of discomfort big enough to be even then perceptible.

After four or five hours' so-called rest we agreed that it was time to be stirring. We were shivering so much that we feared the rotten mountains in the neighbourhood might be upset. There was no talk of eating anything. The packs were made up and I started on. Ten minutes later I was standing at the end of the valley and shouting wildly to Garwood. The object of our journey was accomplished, for the view that greeted my eyes revealed all we had come to see. When he joined me we were able to forget our discomforts together in the delights of discovery.

Before us lay a wide flat valley, known to hunters as the Stordal, but originally named by the Dutch the valley of the Ondiepe or Shallow River, wider by far than and as flat as Advent Vale, stretching away south-west for some miles to Low Sound, and very gently mounting north-east a longer distance before turning a corner with a slow wide curve to the east. Down Low Sound we could see the purple rocks of the Middle Hook that separates Low Sound from Van Keulen's Bay, and farther off still the peaks of the promontory between Schoonhoven and the sea. The Sundewall Mountains were clear of clouds across the flat. Immediately over against us three valleys, each containing a glacier, debouched almost together into the Ondiepe Valley. Farther up, pouring out from the east on to the flat, was the wide, circular-domed front of a vaster glacier, draining a high snowy area of considerable extent. With ponies and sledges we might now have advanced straight up the great valley for many miles, and perhaps come within rushing distance of the east coast. At once the thought occurred to both of us that Bolter Pass, which we had not yet explored, might be practicable for the beasts, and if it were, the thing to do was to bring them over at once. There was nothing to be gained by wandering about on the flat. Our business was an immediate return.

Before setting forth we had one more good look round. The view was worth looking at for all reasons ; to us trebly precious for the price it had cost. No ray of direct sunlight illumined it, no fresh green cheered it, no blue sky overarched it. The brown flat below, just emerged from winter snow and not yet carpeted with its summer inflorescence, spread abroad, imposingly expansive. In my experience all wide views over new country suddenly revealed, produce on the beholder, at first glance, the same exaggerated impression of extent. You feel the size of the thing more than aught else about it. The relative smallness of the snowy mountains, to one accustomed to associate glaciers with high altitudes, added in our case to this effect of breadth. Cold looked the snows, enormous the great glacier snout in front, mysterious the slow bend of the great valley, and the tantalising secret of the hidden regions behind. Could we but climb Stortind on a fine day these would all be revealed. We noted it as a point more than ever worthy of attainment. Garwood took a round of photographs, I of compass bearings; then we turned our backs on the scene and set forward to return.

One thing we agreed upon; we would not follow our old tracks. I went down to the immediate margin of the stream, and found good ice overhanging the torrent, but strong enough to bear. Along this progress was easy. Garwood climbed high up the hill-side to the level of the old moraine, where also he got on well enough. The result was a divergence too wide to be advisable in such a remote region, so I presently climbed up to him, and we journeyed on together. It is needless to recount the details of the way. We had less snow bog to wade through, but more laborious work of other sorts, for we struck the great terminal moraines of the glacier we passed the previous day too high up, and had to mount and descend over them, then to come down just under the snout, reascend the big moraine on the farther side, again descending to a slushy snow flat, where all the worst features of the previous day's struggles were renewed. There was much survey work to be done, and many observations to be made. We were weary, and our backs protested against their loads. Feet, sodden for long hours, became tender. The pace grew slower and slower. About two P.M. we reached Bolter Pass (1340 feet), and saw far away to the north the site of Bolter Camp. We were regretfully driven to acknowledge that ponies could not be dragged over this route till quite late in the season, if then. Possibly it may never be practicable for them. We were too tired to be much disappointed. Indeed, the certainty that this way would not have to be retraced was for the moment a distinct pleasure.

The descent led down waist-deep snow to a large open space, soon revealed as a lake of snow slush, quite impassable. It was turned with difficulty above its head by wading through rivers of a nameless compound neither solid nor liquid, neither ice, water, nor snow, but which possessed the qualities of being as wet as water, as cold as ice, and, whilst offering no support to the tread, opposing a deadly heavy obstruction to the advancing foot. Below the snow-lake was another smaller one, both in the midst of a white wilderness, and below that the floor of Bolter Valley. Two glaciers debouch into it, on a common ice-foot. That from the west leads up to the plateau, and by it a way could be found to the valley of Coles Bay. The other, named Rieper Glacier by Garwood (for purposes of geological reference), drains the north-westerly part of the Fox Plateau. Its great snout was of the normal form in these parts, rounded both vertically and horizontally, and apparently advancing. There were recently deposited beds of boulder-clay before it, which instantly attracted the attention of us both.

The interesting phenomena of the Rieper Glacier's snout, coupled, perhaps, with a sub-consciousness of the horrible nature of the remainder of the walk, conspired to keep us long delaying. When we set forward it was with the determination of getting back to the tent as soon as possible. Doggedly we plodded down the right bank of Bolter Valley, then round the end of it over a boggy shoulder to the sledge tracks of our former way. A pair of ptarmigan close at hand were so well aware of our feeble condition that they let us come within three yards of them. One would have made a welcome dish for supper; I heaved stones at it, but so feebly that it merely hopped aside and let me go on heaving till I was ready to drop. There was less snow by far than when we were here before, but as many bogs and even more water to wade through. Little recked we of wet. Not a word did we speak, but just plodded on, passing the foot of gully after gully, and wondering when the last would come. Two more ptarmigan treated me as the others had done. I knocked some feathers out of one with my ice-axe, but could not catch him. In this country a second-rate archer might have good sport. Birds and reindeer are just tame enough for such a person to stalk with some chance of success.

At eight o'clock the last slope was mounted, and the green tent lay as we left it. Nothing was changed but the wind. A fatal mutation! involving the repitching of the tent, for it would be blown away unless its door were leeward. The pegs, so securely planted, had to be hauled up, every article of baggage to be drawn forth, everything moved and re-arranged. It was half-an-hour's work, performed by both in a silence as complete as it was full of meaning. The cooking apparatus and stores were gathered before the door. Then at last we could doff our wet nether garments, plunge up to the knees in the warm reindeer-skin bags, and take supper in hand. O hot ration-cartridges, red, blue, or of whatever colour! O succulent cocoa, and slowly uncondensing milk, and ye, O Reading biscuits, covered with butter and brown sugar!—long will the memory of you abide in the thankful bosoms of two famished bogfarers. The times of our going to sleep and of our next day's rising were not recorded.

AN INLAND CAMP.

CHAPTER VII

ADVENT VALE TO THE SASSENDAL

IT is perhaps needless to say that the morrow after the laborious expedition recorded in the previous chapter was a day of rest and writing, for there was much to be written up and plotted in. Hour after hour the wind howled and the tent boomed and blustered, but there was no low fog, so that as we sat on the tent floor we could always see the fjord, here steel-grey where ice covered it, there blue or purple in the changing lights, whilst the slopes and bogs that framed it in, now almost bare of snow, displayed a richness of dark tones not easily surpassed. The sight of Advent Point served to remind us how short was the distance inland thus far covered. If the severe labour we had undergone had only brought us thus far, how difficult would be the task of crossing the island even at its narrowest neck!

With evening came a downpour of torrential rain, to the amazement and disgust of Garwood, who had withdrawn to the retirement of his sleeping-bag whilst some of his things were out for an airing. Either they or he must be soused—a miserable alternative. How the rain rattled on the tent,

ON THE WAY TO SASSEN BAY.

driven by the wind! Who, I wonder, told us that rain seldom falls in the summer in Spitsbergen, and then only slightly and with little wind? The same person perhaps who said that there were no rivers in the island! As the hours advanced the storm became ever worse, but it manifested the excellence of our little tent, which neither bowed before the gale nor let the rain through. For their credit let me here record that it was manufactured by the Military Equipment Stores.

In the midst of our first slumber that night, we were aroused to consciousness by a voice without, and, looking forth, beheld Gregory standing alone in the midst of a universe reeking with wet. Water ran off him in cascades from all points. He said the others were near at hand. Trevor-Battye and Pedersen presently appeared with the ponies and lightly-laden sledges, having come straight through from Advent Point, in the very midst and fury of the opposing storm. Trevor-Battye had come to pay a call and discuss plans. He made light of his discomforts and only craved a little food before setting forth on his return, being too wet to remain without bedding or change. He sat in the tent door, patiently eating his ration, and agreed to come round by boat to Sassen Bay in four days' time and meet us there, weather and ice permitting. As a matter of fact weather and ice did not permit. What actually happened was as follows. The gale then blowing carried the heavy East Coast ice up the west coast of Spitsbergen. Next day the fickle wind changed and blew this ice into Ice Fjord, packing the mouth of Advent and Sassen Bays. Day after day the procession of ice-floes continued, and no boat could make head against the wind or find a way through the packed ice to come at Sassen Bay, though various attempts were made. At last on the 6th they were able to leave—Trevor-Battye, H. E. Conway, and a sailor. A day's work carried them through large floating ice blocks to the mouth of De Geer Valley, where they met the Swedes, who had recently shot a polar bear on a floe, and had caught three young foxes on Goose Island. They presently put forth again and rowed on, worming their way with much difficulty and some danger through the ice, and so, at ten A.M. on the 7th, they gained Starvation Bluff, and there met our man Pedersen and established communications with us.

To return, however, to the affairs of the moment. Trevor-Battye and Gregory had brought up the two sledges, but there had been only time to mend one of them. The other was in a parlous state. I would willingly have waited two more days for the work to have been properly done, but Trevor-Battye thought that speed was the essence of our needs, and acted accordingly with the best intentions. The result was unfortunate, but no one was to blame. We had to push on with such machines as were available, and the two new sledges must come round by boat to meet us at Sassen Bay. Trevor-Battye presently said good-bye and went forth into the rain. We tucked Gregory into a bag between us, and Pedersen pitched the second tent close by.

The Advent Bay news was interesting. Mr. Arnold Pike's yacht had come in from the north, with tidings that the year was an unusually open one. The ice had disappeared to a high latitude; the sea was open as far as the Seven Islands and perhaps yet farther east. A school of white whales had visited Advent Point, and Studley shot four of them and felt better. Svensen's walrus-sloop also came in and took Studley and Trevor-Battye for a cruise over to Cape Boheman. The day we climbed Fox Peak, Gregory ascended the bluff above Advent Point and had a glorious view.

When we turned out the rain had passed, and the cloud-mantle resumed its usual spread and level. I packed one

ADVENT VALE FROM CAIRN CAMP.

sledge and set forth with the plane-table, the others following an hour later, Pedersen driving one sledge and Garwood the second. Our party, thus constituted, was an ideal one. Gregory was quite well again, and gave us the benefit of his boundless energy, his alert observation, and his wide experience. Garwood was not a whit behind him in every virtue a traveller can possess. Henceforward, with a self-sacrifice I can never forget, he undertook the abominable work of driving one of the sledges, so that I might be always free to go ahead, aside, or to linger behind for surveying purposes. It was Garwood, too, who did the cooking for us, and a thousand and one other details of camp-work that would have been less efficiently done by any one else. Pedersen was of little use in camp, but, when we afterwards exchanged him for the other man, we were even less well served.

On descending from camp to the plain, the many streams from Fox Valley had to be waded immediately. An almost level stony area followed, the best walking we found in Spitsbergen, for the ground was firm and fairly smooth. Below the wide long fan came the flat land, occupying the bed of what was once Advent Bay. This I traversed to its head, where is still the old bank curving all around. I climbed a hump, previously selected for plane-table station, and gazed abroad on the dismal prospect. Clouds, of course, disguised the peaks, but here and there a point was visible through some momentary gap, so that useful work could be accomplished. The others were far away like tiny ants working at the sledges, which began to move just as I set forward once more.

The upper part of Advent Vale was soon to be revealed to me, and expectation stimulated advance. On the left flowed the river, no inconsiderable flood, over the flat bottom of a gorge whence long bog-slopes rose on either hand into cloud. A fold of the ground hid all the area that had become so familiar to us during the last few days. I was absolutely alone in a new world, hitherto seen only by occasional reindeer hunters. Under such circumstances the dullest imagination should be quickened. Nowhere was there trace of man. Abundance of reindeer horns lay about, and the tracks of reindeer were in all directions, but not a footprint of man or of any domestic animal could be descried. Wild birds flew about me inquisitively, flocks of geese honked at me as I moved forward. The scenery was, in fact, tame and dull, but circumstances invested it with a strange prestige. Its rich purple tones, its wide-expanding forms, its suggestive peeps of cloud-enveloped crags, sufficed to quicken the fancy, so that I walked along in the bleak dull day as in a dream, full of a nameless and indescribable delight. Work involved frequent halts, for the bend of the valley closed out known points and opened new ones that had to be linked to the old before the opportunity was lost. Thus the others came up with me in course of time, all wet to the knees like myself, and we could indulge in a little common complaint. The ponies were tired after their long march the day before, and they too complained, though the route had continued unusually good.

A mound of ancient moraine here stood in the midst of the valley. Evidently it must have been formed by a glacier, which once descended the large side-valley from the north, whose mouth we presently opened. If the clouds had been less thick and low we should doubtless have noticed the importance of this depression, but it was only a month later, when we looked down upon its head from the top of Mount Lusitania, that we discovered it to be the end of a deep trough, cutting through the mountains from Ice Fjord, and dividing into two almost equal parts the mountain group between Advent Vale, the Sassendal, and the fjord. Any one ascending De Geer Valley, and crossing the low pass at its head, would come out at this point into the upper part of Advent Vale.

When the valley opened out once more there were gullies to be crossed, some wide and with steep or even precipitous sides. Long detours and much uphill work was involved, we fell into bogs and snow-beds again, and progress was difficult and slow. At last a gully, bigger than all the others, caused protracted delay and labour. When it was crossed we had to halt for the ponies' sake, though the position was abominable for a camp, and the march had been short. It was one A.M. There was not a yard of dry ground, neither was there any clear water. The earth was formed of dirty black débris, damply compacted with moss into an oozy bog, which spread afar over the sloping face of the hill. One spot was about as bad as another, so we halted near a muddy trickle of water and pitched Black Ooze Camp (530 feet) on spongy ground, that squdged wherever we trod, and quickly evolved puddles from our footprints. On the opposite side of the valley reindeer were visible, so after supper Pedersen went forth for meat. He saw about a score during a short stalk, and brought down two of them—a welcome addition to our supplies, though one would have sufficed. The ordinary business of camp closed the day.

We had by this time settled down into camp-life routine. At first the smallness of the tents was a constant annoyance, and we were used neither to the sledges nor to the cooking apparatus. Now we, or rather Garwood, cooked with promptitude, not to say elaboration. Variations upon the sameness of the earlier *menu* were devised. Reindeer and ptarmigan ousted ration-cartridges from the pot. We grew fat instead of lean, and only the ponies suffered from short commons. Camps were pitched and struck with rapidity;

sledges almost loaded themselves and required far fewer ropes to bind them than at first. The open-air life invigorated our frames; hearts beat strongly, and we could defy cold with impunity, neither did we dread the constant wettings we had to undergo. We even became used to the bad weather, and ceased to repine that a blanket of cloud kept sunshine from us save at rare intervals.

At four P.M., July 1, I left the oozy and uncomfortable camp and climbed a rounded and unattractive knoll (834 feet) behind it, a mass of bare débris, standing in the midst of the valley, and likely to command a serviceable view for topographical purposes. Cold blew the wind over the level top. Clouds covered all the larger hills, but I was able to look far up the valley to its final branching, and to mark the situation of two cols leading over into the Sassendal. Gregory and Garwood presently followed, their cries revealing from afar that they had discovered fossils. Frozen to the bones, when my own work was done, I joined them, and, borrowing Gregory's hammer, went to work stone-breaking to warm myself. The fossils came forth by dozens. It was as amusing as catching fish, though in this also experts had the best of it. I smashed away, and found only bad specimens, whilst rare treasures leapt forth to their lightest taps.

Leaving them thus well employed, I returned to strike my tent and pack a sledge, then set forward up the main valley, not as before across the bog slopes, but down on the flat stony bottom of the shallow river gorge.[1] Knowing how many times water would have to be crossed by this route, I foolishly wore long rubber boots, which kept the feet dry indeed, but drew them to lameness. It was possible now to advance rapidly, for there were either reasonably

[1] The altitude of the river-bed below Ooze Camp was 327 ft. above sea level.

THE BALDHEAD FROM BRENT PASS.

STUCK IN A SNOW-BOG.

hard stone-flats or equally level marginal ice-beds to go on. Good progress was therefore made, and in an hour I was standing where the valley broadens out to a great basin, and a long wide view can be obtained. I climbed the bank and gazed around with real delight. Far away in the sombre purple distance were the hills behind Advent Bay. In the opposite direction, across a mile of grey stone-flat, cut up by scores of streams (all to be crossed by us), lay our pass to the Sassendal, clearly visible and of easy access—a wide low opening in the hills. The main valley wound gently up to the SW. between hill-sides striped with melting snow-beds. But the striking feature of the view was farther round to the south, where the huge cliff-fronted and serac'd snout of a glacier swept forth from a deep blue cloud-enveloped valley. The ice-cliff, appearing thus literally from the clouds, produced a most impressive effect. It thundered forth at frequent intervals, when masses of ice fell from it, so we named it Booming Glacier. Gregory, who came up at this moment, was so taken by it that he set out for a nearer view, whilst I levelled the plane-table to secure some observations before the dark storm rushing up the valley rendered work impossible.

When Garwood came by with the sledges, we began crossing the streams. The mended sledge went bravely through, but we watched the other in fear and trembling, for one of its runners was worn almost as thin as paper and evidently would not hold out much longer. It waved and twisted over and about the stones, but for this time also did not break. At the far side was a bank of snow-bog to be climbed, wherein the ponies floundered up to the withers. Spitz set knowingly to work, prodding with her fore-feet to beat down the snow, and when she found it too deep, refusing to advance till I had gone before and trodden a broad deep furrow. Thus we came safe to land on a dry bank almost level with the col, which we named Brent Pass, from the flocks of geese that kept flying over it.

The ponies arrived fairly famished, for there had been no grazing worth mention at Ooze Camp, and we were carrying no hay. Thus far Bergen had been cock of the caravan. Now, over their feed of oats, Spits resisted his supremacy, and a fine squabble arose. Kicks were freely exchanged, and what sounded like equine abuse. Before they could be interfered with, Bergen so definitely got the worst of it that he always knuckled under in future, letting Spits deprive him of the remainder of his oats, or any succulent morsel she might find him devouring, if no one was by to see fair-play.

The geologists went to investigate Booming Glacier next day, and found its sides as steep and almost as inaccessible as its front. The ice-stream bulges and cracks all round its edge and up both sides apparently for miles. With difficulty and some danger they climbed on to it, returning even more learned than they went. So fascinated was Garwood with what he saw that he returned three weeks later to Brent Pass, and climbed the peak (2868 feet) above the left bank of the glacier, which in its turn is separated by another glacier from the wider mountain that rose immediately opposite to our camp, to which we gave the name Baldhead. The group of peaks that surround Booming Glacier is the culminating mountain mass between Advent Vale and the valley of the Shallow River. If it had not been for our appointment to meet Trevor-Battye at Sassen Bay, we should have spent two days here for the purpose of exploring the head of the glacier.

I spent the day on the opposite slopes. Two reindeer at one time visited me, looking, running away, returning, running away once more, and so on for half-an-hour. Later, I walked over Brent Pass (450 feet) to discover the best way

THE SNOUT OF BOOMING GLACIER—BRENT PASS IN THE DISTANCE.

for the ponies, and decided on following the right bank. The actual pass is wide and swampy. A stream, descending from the hill on the north, divides and sends one branch down Advent Vale, the other to the Sassendal, clearly an unstable arrangement. Formerly this stream went wholly down Advent Vale. Soon all its waters will have been robbed for the Sassendal by the backward creeping Esker Valley. The watershed between Advent Vale and the Sassendal used to be close to the left bank of the latter. The Esker Valley was one, though not the longest, of the upper branches of Advent Vale. In a manner to be presently described the old head of the Esker Valley was cut down and became its mouth, and the slope of the valley has gradually been turned the other way. The process of eating back is steadily continuing. Brent Pass will be eaten away, and the Advent River will suffer a further diminution of its head-waters, for its main upper tributary will be taken from it, and the pass will then be situated at the foot of the Baldhead. The process may even go farther, and the ultimate position of the watershed may be at the narrow place just above what was formerly the head of Advent Bay.

There was no distant view from the pass because of clouds, which hung low in all directions, yet the scenery was invested with no little dignity by its long sweeping lines, its simple forms and sombre colouring, seen through the thick wet air. The broad bases of the hills seemed to imply vast mountains rising from them within the clouds; these the imagination easily supplied. How an ancient Greek poet might have filled such a scene with gods and mighty heroes aloof from little men! Across such country Hercules might stalk at large and seem in place, or hither might come mysterious divinities not to be approached or beheld of human eye save after awful initiations. Alas for the age of science! The hills were a topographical puzzle for us, and their story a geological enigma.

In the afternoon of July 3 we awoke to find the sun pallidly shining on the bleak hills, through a sky beautiful with various layers of cloud up to a serene bed of delicate cirrus. Cold draughts blew over the pass, and there was a thick cloud-bank beyond, but the omens were at last fair, and Baldhead Peak was mainly uncovered, whilst up Booming Glacier were tantalising peeps of deep-lying snow-fields and other perplexing forms. The first work was to patch up the sledges, one of which was in a parlous state, its runners reduced to mere paper, hopelessly frail. Whilst this was in hand—or, to be accurate, in Gregory's skilful hands—glaucous gulls flew around, and inquisitive long-tailed skuas hovered only a few feet overhead. Food had been found for the ponies, who were thankfully absorbing it, after two days' uncommonly short commons, or rolling on the moss and mud in joyous relaxation.

Encouraged by the opening prospects, I set forward to work, crossing the pass, and almost immediately losing sight of Advent Vale. The scenery was utterly tame, bogs below, then gentle rounded slopes leading up into clouds, which, however, hid no crags or diversified sky-line, as we afterwards discovered. Only the Baldhead kept showing behind, as though peering to see us safely out of his solitude. Five grating geese kept in attendance; their throats needed oiling. Then two reindeer came and looked at me, and being puzzled came nearer to within about forty yards. They were asinine-looking beasts, with an awkward waddle in their going and a grunt for voice. Finding me dull, they fed awhile, then made for the river. The snow-bog bank was rotten. The leader, sinking into it up to the neck, returned and tried another place, where was an ice-cornice, which broke under him, and let him in up to the neck again,

LOOKING DOWN THE SASSENDAL.

this time in the river. Between swimming and wading both crossed, shook themselves like dogs on the farther bank, and trotted off up hill. These reindeer have been shot enough to be made shy, but they are essentially stupid beasts, and if Spitsbergen is much hunted they will soon be almost exterminated. I saw many more, and so, unfortunately, did Pedersen, who cannot restrain himself from shooting at them, even though they are not wanted for food. He is not a good shot, and this day he was specially unfortunate. Having stalked two deer and come within forty yards of them, he kneeled down and elaborately aimed. The recoil of the rifle sent him over on his back, and no harm was done; the stupid deer wondered at the noise, and came close to see the cause. Pedersen missed them again, but brought one down by a long shot at a range of about two hundred yards—a useless slaughter, as will be seen.

My march was a solitary one down the mournful Esker Valley, but it was not without interest. The very loneliness and weirdness of the scenery gave it a charm. Had Sir Palamedes lit on this place, assuredly he would have believed that the Questing Beast was nigh. On I went, plunging and wading slowly downward. Perceiving that the exit into the Sassendal was near at hand, I hastened forward, ascended a bare and muddy mound, and lo! the broad valley, and to the north a bright line, the ice-encumbered waters of Sassen Bay. Across the valley was the Colorado Berg, a long low hill, extraordinarily flat-topped and tame, seamed by snow couloirs, one of which, with its branches, mimicked the form of an Assyrian bas-relief. Here was clearly the place for camp if the ponies had come down this bank; there was nothing to be gained by descending the last low boggy slope to the yet boggier flat of the valley. Long I sat gazing at the view and shivering. It was not without its fine elements. Best was the glimpse of Sassen Bay, with the purple wall

H

of Temple Mountain beyond. The head of the Sassendal, of course, attracted my most anxious attention, for thither lay the route to the East Coast, a portion of which was traversed by M. Rabot in 1892. The bluff he climbed and named Pic Milne-Edwards was in sight, and so was a wide and gentle glacier, stretching back eastward with slow ascent to a flat white sky-line. I named it Rabot Glacier.

In the faint sounds that fluttered the air it was surprising to detect a deep bass note, giving volume to the treble ripplings of brooks down the hill-sides, and of the main stream hurrying over stones or washing against its snowy banks. A brief search revealed the cause. The river at its exit from Esker Valley passes through a gorge, cut into the carboniferous limestone which here forms the foundation of the triassic hills. At one point it has to tumble over a wall of rock about fifty feet high. Over it goes at a level edge, in a single plunge, a plain unbroken fall of brown water, straight sided, foam footed, a cascade in its simplest form. It tumbles into a wall-sided caldron, and winds away below a few jutting promontories, which form excellent points of view, and were chosen for nesting-places by a colony of pink-footed geese. The scene is admirably harmonious in its dreary simplicity and sombre remoteness ; even the muddiness of the water is better in keeping with the surroundings than clear water would be. It is a small waterfall ; later in the season when the snow had done melting it would be smaller, but it was my own. No one had ever seen it before, except perhaps some stray hunter. It was mine to name and to enjoy, which I did with trembling, for the wind was cold and there was no shelter.

The hours passed and no one came. Knowing that the march would be a hard one for the sledges, I now began to fear lest some misfortune had happened. The reports of Pedersen's rifle at last relieved me. Then Gregory came in

CAULDRON WATERFALL.

with tales of the difficulties surpassed—how Garwood's pony had been bogged and the other had to be brought to haul him out; how the sledges were only just holding together, how in descending the right bank they were forced into a *cul de sac*, with a high bank overhanging the river on one side, and a slope of snow too steep to be traversed on the other; how in fording the river the sledges were rolled over and carried away, and they themselves with difficulty retained their footing, and how they were all dripping wet and as hungry as I was. After midnight the caravan arrived, and Waterfall Camp (105 feet) was pitched on a dry spot commanding a wide view.

DR. GREGORY.

ON THE TOP OF STICKY KEEP.

CHAPTER VIII

THE ASCENT OF STICKY KEEP

SUN brightly shining on the tent, and air, in consequence, stuffy within, were the unarctic conditions that aroused us from slumber, early, but not early enough to prevent the headache which sunshine on a closed tent is so liable to cause. Perhaps this was why all felt lazy. The attack soon passed off, and by two o'clock P.M. we were ready for our tasks. Gregory went off with the wrecked sledges, expecting to find Battye and a boat at Sassen Bay. If all went well he was to load the ruined sledges into the boat and take them to Advent Bay, to be sent thence to Tromsö or Hammerfest for repairs. He was to return at once with the other two sledges that had not yet been used, and these, it was hoped, would take our things safely across the island to Agardh Bay. Delay being thus imposed upon us, it

Printed Lemercier, Paris

BOOMING GLACIER, THE BALDHEAD, AND FOX PEAK FROM STICKY KEEP.

was not surprising that fine weather should at last set in. Pedersen went down with Gregory, and Garwood and I were left alone.

The obvious thing for us to do was to climb the hill behind camp. It has been said that the opposite or east side of the Sassendal is bordered by the remarkably uniform face of the Colorado Hills, which is only broken by one or two insignificant valleys, or rather cañons. The west side of the Sassendal, on the contrary, is entered by a whole series of side valleys of considerable size. Delta Valley, the nearest to Sassen Bay, gives access to Advent Vale by a pass which descends almost on to the top of Brent Pass. The second side valley is the Esker, at whose mouth we were encamped. The third leads by a pass to the upper branch of Advent Vale. The fourth is Fulmar Valley, by which we went to the Ivory Gate, and Agardh Bay. Between each of these valleys and its neighbour is a jutting mountain front or bluff. The first of these is Mount Marmier, above Sassen Bay; the second is the bluff we climbed and named Sticky Keep; the third protrudes three bluffs towards the Sassendal, and was named by us the Trident; the fourth is the bluff climbed by M. Rabot, and named Mount Milne-Edwards.

We set forth to ascend Sticky Keep, Garwood taking hammer and camera, I the everlasting plane-table and its irritating and needlessly bulky legs. There was no climbing on the peak, but a great deal of miscellaneous steep uphill walking, all toilsome and disgusting in various ways. To begin was a slope of smooth hardish mud, about 1100 feet high, a featureless steep incline, seamed with a few shallow snow gullies, but otherwise the same from side to side and from top to bottom. On this Garwood and I parted company. Above it came a large plateau, as it were paved with flagstones falling into decay, the very semblance of some ancient ruin, which it was hard to believe the mere work of nature both in structure and decay. The peak was seen rising beyond in a series of steps, like the keep of a castle within the *enceinte*, or like the storeys of some Chaldæan *Ziggurat*, rotting back to their original mud by the banks of Tigris or Euphrates. The way led over sticky mud, into which one sank ankle-deep, the mud flowing over the foot and adhering to the boot, dragging it back both by weight and suction. Other paved parapets succeeded and other staged ascents, over slopes of broken stone so steep that it was marvellous they did not fall in one great avalanche. The mud that compacted them made the steepness of the slope possible. The foot behind slid back as the other was raised. It was a toilsome task to go on advancing, and withal a thirsty, for there was little water by the way, and all of it muddy. At last the fifth shoulder was passed and the summit gained (2185 feet)—not a peak, but an undulating area of an acre or more, part snow-covered, part like a ploughed clay-field after heavy rain.

North-eastward, a cliff fell away to the Sassendal; on the other side were long slopes of snow and a round-backed ridge, trending toward Brent Pass. It was a poor mountain, but the view from it was by no means poor, and in the clear sunlight, below the soft blue sky whisked over with cirrous clouds, the snowy regions all around shone through the mellowing atmosphere, glorious in the splendour of the silver world. Far below spread Ice Fjord, with Sassen Bay in front, and the mouths of Temple, Klass Billen, and Nord Bays all discernible. Blue were the waters, with threads and areas of white ice interlacing over them, and keeping the surface so calm that the mountains beyond were clearly mirrored back. Bluer still, incredibly blue, were the dark rocks of Temple Mountain and the cliff-fronted neighbouring peaks. In the warmer distance were the great glaciers of Cape Boheman and the maze of peaks

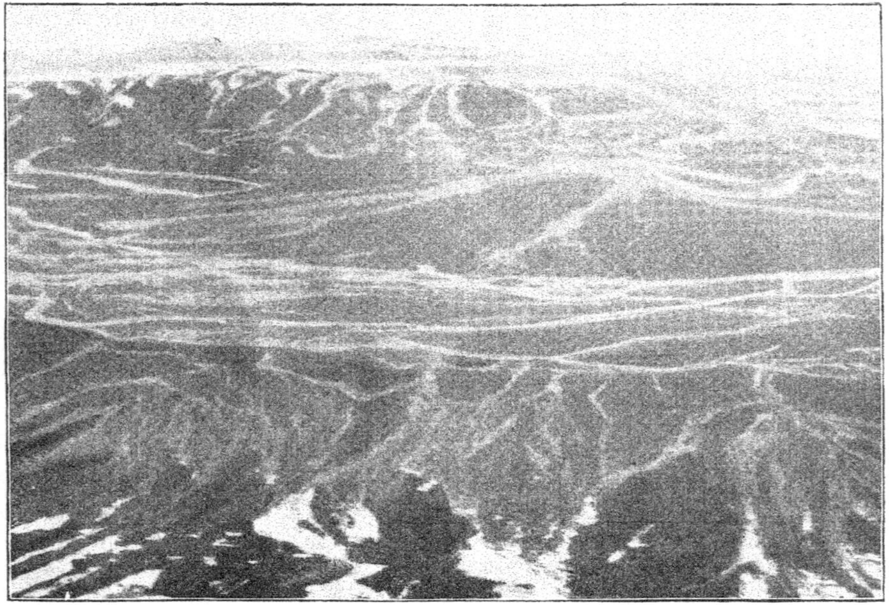

FROM STICKY KEEP LOOKING ACROSS THE SASSENDAL TO THE COLORADO PLATEAU.

which divide them; Mount Marmier, near at hand, cut off the remote panorama. Of special interest to me were the collection of mountains from which Post Glacier descends to Temple Bay. They are marked on the chart like *nunataks*, peaks rising out of a sheet of inland ice, just

STICKY KEEP AND THE SASSENDAL FROM GRIT RIDGE.

what we desired to see; but they are nothing of the sort—only a collection of points and ridges of the usual type; feeding, with the snows they gather, a series of glaciers whose direction of flow is determined by their form and position. In the neighbourhood of Temple Bay is no ice-sheet in the Greenland sense of the term, but rather a miniature group of Alps. The brown flat Sassendal stretched abroad at our feet, below a cliff, whose black buttress-knees jutted out against it, each pair with a snow couloir between. The nature of its floor was now as plainly evident, where were bogs and where dry places, as though it were a damp piece of stuff held in the hand. The lower slopes were beautifully decorated with an intricate tracery of stream-furrows, like skeleton leaves, whilst down the midst, bending in wide curves, ran the purple and steel band of the main river, and its stony and changeful bed, ending a few miles below Rabot Glacier, in an area of snow-bog and ice-foot, apparent from this distance as a smooth white plain. The glacier itself stretched far back in unbroken sweep to a level white sky-line, but there were indications which made me suspect (what afterwards proved to be the case) that we were really looking over a col, and that the remoter snowfield is drained away down the other side of the watershed. It, in fact, feeds the ice-tongue that fills the head of Mohn Bay.

Over against our peak, along the far side of the Sassendal, stretched the curious assemblage of the Colorado Hills. The nature of this area was now most interestingly apparent. It is a portion of the old plateau, which, till recently, was protected from denudation by an ice-sheet. This has been withdrawn, and the surface, no longer protected, is being cut down by the action of running water. A series of cañons is being formed, from which the region was named by Nathorst in 1882. The ice-sheet once spread far to the west, but long ago retreated up to the edge of what is now the Sassendal, so that the region through which we had recently come is occupied by mountains which are of a more developed and less rounded character the farther west you go towards Advent Bay. It was only some weeks later that we were enabled,

LOOKING SOUTH FROM THE TOP OF STICKY KEEP.

by a closer examination, to establish this interesting fact. The Colorado region shows the first beginnings of hill formation by water action. The row of hills on the south-west side of the Sassendal are typical of the next stage of development. The hills between De Geer Valley and Advent Bay are complete and developed mountains, with sharp ridges and forms of marked individuality.

Beyond the Colorado Hills a range of higher peaks stood up against the horizon, but only their summits were visible, and we could not discover the principle of their arrangement. They appeared to be connected with the Temple Mountain group, and to enclose glaciers draining down to the Post Glacier and so to Temple Bay. This is a region that would well repay exploration, and would be approached most easily by way of the Post Glacier.

Turning our backs on the Colorado Hills, we had before us the region through which we had come, when clouds permitted us to see no more than the bases of mountains whose whole mass was now revealed. Booming Glacier stretched back prominently in the midst of a tumultuous region of hills. It is a more important ice-river than we supposed, and flows down in sinuous curves from a remote snowfield, about and beyond whose head appeared a tantalising multitude of peaks. Between the Baldhead and Fox Peak again, was another considerable glacier basin, of whose extent we unfortunately never attained accurate information. The view, as a whole, was of a region in which man has no abiding-place—a land not made for man, but mainly inimical to him. In such a world the human species would swiftly degenerate and presently disappear. Birds and reindeer would alone survive, and the highest civilisation would be that of the glaucous gulls.

Garwood joined me on the summit, when the plane-tabling was done. Together we went again round the view, photographing, discussing, and agreeing upon conclusions. The wind was cool, but the sun so warm that, in descending, we found it agreeable to rest in the shade below each shoulder of the ridge. The sliding débris was pleasant to descend. Then came a snow couloir and a short glissade below it. On reaching camp I took my rifle—for we had no shot-gun—and went to stalk a company of pink-footed geese who had come to visit the twelve nesting couples by the waterfall. One, I thought, was as good as in the pot, but the rifle missed fire, thanks to the state of rust into which Pedersen had allowed it to come—small blame to him, considering through how much water the sledges had been dragged.

After supper came a golden midnight, when the sun shone warmly and the sky was clear. Between wind and waterfall, sounds reached us through all the sleeping hours as of trains going into tunnels and cuttings. Long and late we slept, making a "Europe morning" and a day of rest of the Sunday (July 5) that followed. It was a day amongst a thousand, worth winning by weeks of labour and wet. Cool airs played around; the sun was warm, and the pale blue sky brilliantly clear. It might have been an English May day. There was a Sunday sentiment in the air. One almost expected to hear church bells pealing from afar off over the russet Sassendal flats, but the only sounds were the booming river, the rippling brook, and the flapping of the tents in the breeze.

Two reindeer came and stalked us. They wandered up to the tents and smelt at the sleeping-bags lying out for an airing. We took snapshots at them with harmless cameras till we were tired, and left them to amuse themselves as they liked best—one might as well shoot sheep in a field as such careless and stupid beasts.

The snowfield at the head of Rabot Glacier was absolutely clear, and, to an unobservant eye, might have seemed

REINDEER.

near at hand. Its névé was grey rather than white, a characteristic appearance in these regions, if the always low sun happens to be at the back of the observer, so that little light is reflected to the eye from the almost horizontal surface of the snow-field. In the Alps névés are seldom so horizontal, unless they occupy depressions, where the low sun only casts shadows upon them; moreover, in lower latitudes the sun is much nearer the zenith most of the time, and when it descends to the midday level of Spitsbergen its light comes through dustier air beds, and is thereby deprived of the whiteness which even the low Spitsbergen sun generally retains. Owing to the relative gentleness of winds here, and to the larger proportion of snow-covering at high levels to bare rock, névés in Spitsbergen are cleaner, carry less surface dust, than they do in Alps or Himalaya. The whiter surface and the cleaner low levels of atmosphere thus conspire to deprive distant snow-fields of the more varied colouring we are accustomed to in Europe. Only at distances of twenty miles or more is enough of the clear atmosphere interposed to cast a mellow glamour over wide snowy prospects in the Arctic regions.

About seven P.M. Gregory walked into camp, calling aloud for tea and victuals. He was bearer of no good tidings. He told how, after leaving us the day before, the tired ponies dragged the light sledges slowly along the hill-side, over boggy places, towards ever-receding Sassen Bay. They passed the mouth of the first valley (into which we looked from Sticky Keep), getting fairly ducked in the stream. A while after, Gregory found the biscuits in his pocket sopping wet, and set them on the sledge end to dry. It was the last he saw of them. Spits lets no food remain long untried within reach of her omnivorous maw, and she already knew by experience that biscuits were good. The next stage of the way was when they rounded the foot of

Mount Marmier, and turned along the shore of Sassen Bay, where another stream had to be crossed, beyond which they ascended the low Starvation Bluff and halted. There was no boat in sight; our companions had not arrived from Advent Point. Here was a pretty mess! Gregory and Pedersen had practically no food—only a fragment or two of Emergency Food and chocolate. Pedersen was provided with his sleeping-bag, so in that respect he was well enough off, but Gregory had nothing, and was wet to the skin into the bargain. They laid out their things to dry in the sun, and tried to shoot eider-ducks, but without success. Pedersen took the rifle and went after reindeer, whilst Gregory hammered for fossils. No meat resulted. Pedersen missed an easy shot, and said that the Paradox gun was no use. They turned in without eating anything, or rather Pedersen turned in; Gregory had nothing into which to turn. About three in the morning they heard shots fired some way off. Gregory sent Pedersen to find the shooters, and then availed himself of the vacant sleeping-bag and had a short rest. At ten o'clock Pedersen returned with the following note:—

"SOUTH SIDE OF SASSEN BAY,
5th of July '96.

"DEAR DR. GREGORY,—By your man I have heard that you are somewhat short of food, and that you expected to meet your boat in Sassen Bay. Though we have ourselves not much more than necessary, having stopped here somewhat longer than we intended, I venture to send you some simple provisions and matches, so that you will be able to reach your camp and your comrades. In the outer part of Sassen Bay there is tolerably much drift ice, and it is possible that your boat could not pass or did not venture into the ice. It is to be hoped that the ice soon will spread; if not, you may have to reach Advent Bay the same way you came. If we

WATERFALL CAMP AND THE SASSENDAL.

can get out to Advent Bay we will ask your man to start with the boat at once and bring you assistance.—Yours very sincerely,

"GERARD DE GEER."

With this kind and timely relief Gregory was able to await events. He stayed to geologise and watch the effect of the tide upon the ice. Pedersen thought that a boat might have entered the bay on the previous day, and prophesied that in a few hours, with the wind in the quarter it was, the ice would open again. After four in the afternoon Gregory started away, leaving Pedersen with all the provisions, to await the coming of Trevor-Battye. He made a rapid march and arrived fairly done up. It is not the distances that are fatiguing in these parts, but the labour involved in taking every step.

A council was held on our sledgeless condition, and it was decided to feed Gregory up and give him a good rest, to the end that, on the morrow, he should walk back over Brent Pass, and set things moving at Advent Point, whilst we pushed camp across the flooded Esker River and explored the neighbouring peaks and valleys. The delay was less annoying than it might have been, for there was plenty to do in the surrounding country. Now that the short northern summer had set in, flowers were opening on all sides, and several grasses put forth their tender shoots. A veritable Arctic garden surrounded the tents, for the ground was gay with blossom. There were large patches of *Saxifraga oppositifolia* scattered about like crimson rugs. *Dryas octopetala* and the Arctic poppy were as common as buttercups and daisies in an English meadow. Yellow Potentillas (*verna* and *multifida*) added their welcome note of bright colour. The Alpine Cerastium was the gracefullest blossom of the company. Then there were two Drabas, a Silene, *Lychnis apetala*, *Oxyria reniformis*, and a number of other plants not yet in flower, besides the mosses. It was strange to meet again in this remote region so many plants that I had found by the glaciers and amongst the crags of the Karakoram-Himalaya. *Papaver nudicale* and *Saxifraga flagellaris* recalled a wonderful day's march in 1892 up the left bank of the Hispar Glacier, noblest of Asiatic ice-streams. *Lychnis apetala* grew commonly by the great Baltoro Glacier in full view of mighty Masherbrum. *Potentilla multifida* was common at lower levels in Hunza and Nagar. *Saxifraga oppositifolia* and *Saxifraga herculus* climb to a height of 17,000 feet and more on the sides of the greatest giants of that most wonderful range. Here they all were again, as bright, and maintaining themselves as happily in the heart of the Arctic regions as on the backbone of Asia.

The nesting geese by the waterfall gave us constant entertainment, unfortunately for them. We thought it was our visits that made them shy, but we were only unwillingly at fault. It was the glaucous gulls, attracted by our camp refuse, who were the real sinners. When they had devoured our leavings they turned their attentions to the nests, eating the eggs one by one and then the fledglings, till not one remained, and the bereaved geese deserted the place. One day I observed four gulls, in solemn conclave, watching me as I cut joints off a reindeer. The birds saw the joints put into a stream of icy water below our snow-patch, and had a great deal to say about them. Fearing what might happen, I knocked the tail-feathers out of one gull with a bullet, but the lesson did not suffice, for on returning to camp that evening we found our meat gone and two birds a hundred yards off sitting by the bones and chuckling at us. Each joint must have weighed as much as a gull; how they managed to carry them away, without leaving a footprint in the soft bank of our brook, was and remained a mystery.

LOOKING UP THE SASSENDAL.

AFTER THE DAY'S MARCH.

CHAPTER IX

ASCENT OF GRIT RIDGE

AT 5.30 P.M. (July 6) Gregory started for his thirty-mile bog-tramp to Advent Bay. He went forth in the gayest fashion, saying it was some time since he had walked fifty miles at a stretch, but that he thought this thirty might be counted as an equivalent, which indeed was true. The Trident fired a salute of falling stones in his honour. Things having been set in order in and about the tents, films changed in the cameras, and lunch packed, Garwood and I also left, at seven P.M., for a mountain scramble. Our plan was to descend the main valley to the mouth of the first or Delta Valley, which penetrates the high land just beyond Sticky Keep. We mounted gradually over the foot-slopes of that hill, thus gaining an ever wider view towards Sassen Bay, whither we looked in vain for traces of our over-due companions. The bay was practically clear of ice, save a few isolated floes, and its waters were so incredibly calm that the Sentinel Rock and all the details of the Temple's façade were repeated below. It was a day of wonderful colour;

the rich atmosphere dyed rocks and distances as with the priceless product of Tyrrhenian seas, and the eye, gaze in what direction it might, beheld no form that was not largely dignified, and no tints that were not rich and harmoniously combined. A cloud-bank lay flat below the northern sky, and a few collar clouds encircled the throats of nearer peaks, whilst round white islands drifted over the blue heavens; but they formed only to fade again, and indicated no lurking ill-temper of the sky.

An hour's walk led to the north foot of Sticky Keep, which was rounded some 200 feet above the plain. On the wide ridge was a polygonal bog of strange regularity. The muddy disintegration of the friable rocks above, here lies almost flat for a while in its slow, glacier-like descent. Expanding and contracting between extremes of temperature, it becomes cracked up hexagonally. Seeds find lodgment in the cracks, which thus become outlined by ridges of mossy and other increasing vegetable growths. The naked mud within the hexagons again, by processes of freezing and thawing, is forced up into rough and broken domes, whilst the vegetation grows into bosses and clogs the water washing down from snow-beds above. The ground becomes saturated, and a regular bog is formed, which moves slowly downward, sometimes splitting into crevasses, or giving way into stream channels, whilst, at the time of great snow melting, small mud rivers flow down on to it, and further diversify and confuse the soft and tangled surface. As the ground beneath is released from its winter bondage, the whole becomes softer and less stable. Thus it flows, rather than falls, to the valley below, where it comes to relative repose on the wide and ever-widening boggy flat.

Farther up the hill-sides one can watch the process, whereby the mud is formed out of the soft horizontally-bedded rocks of which the hills are built. Melting snow and splitting frost loosen and crack the slabs into a mere slope of débris, all rotten and ready again to split and resplit. Sodden snow and trickling water rapidly produce immense quantities of mud, which partly support, partly carry down the stones that are not reduced to powder. On hot or wet days streams of this stuff, small mud avalanches, in which the mud predominates over the stones, trickle down the slopes, and either reform on gentler slopes below, or flow on to the surface of some glacier. In the latter case they are carried down, and ultimately either shovelled off the sides and end to form astonishingly large masses of boulder-clay, or dumped into the stream and carried straight away to fill the estuary at the river's mouth.

This day was really hot. The sun shone with sensible power, and all the land felt its force. The snows had now been subjected to its dissolving action without cessation for all the four-and-twenty hours of three consecutive days. What melting had thus been accomplished may be readily imagined. One could see the snow-beds growing smaller. Whole white hill-sides became brown between sleeping and waking. Brown slopes with no more snow above them dried into grey cakes. Bog-slopes facing the sun steamed like hot potatoes. Every water-course was swollen to its full capacity, and the brooklets of the previous week were channels difficult to wade. We discovered this to our cost when within Delta Valley its river had to be crossed. It was a raging torrent between ice-banks which it undermined. For some distance there was no point where a passage could even be attempted. A division in the valley was reached, where the stream unites two almost equal branches, one coming from the col leading to Brent Pass, the other from a glacier basin between Mounts Lusitania and Marmier, the exploration of which was the day's aim. The first branch had to be crossed; it seemed no shallower than the united torrents.

Selecting the best place to be found, we plunged through and scrambled out on to the overhanging ice-bank beyond, wet to the belts.

The ice-foot of the Grit Ridge Glacier reached to the junction of the two torrents, so that the second torrent, which was in fact the glacier stream, need not have been crossed at all. Its left bank, however, offered a more comfortable route, so we jumped across at a point where the overhanging ice-banks reached out towards one another. The torrent, with its floor and walls of purest ice and its dark waters, was a beautiful thing. Garwood lingered behind to investigate the structure of the ice, where, at one point, its rod-like crystalline structure was displayed. He almost lost his life in consequence. The corniced bank gave way beneath his feet as he approached the edge to take a photograph. By a fortunate chance he did not fall into the race of waters, whence he could not possibly have emerged alive, for the floor was ice, the torrent was in flood, and the walls overhung like a tunnel. He was facing the stream, and he went straight down, but his elbows behind his back caught on the newly-broken edge, and there he hung suspended, unable to get any purchase with his feet, for they went right back against the slippery and still overhanging wall. He believes he remained in this dreadful position for ten minutes, before, by some twisting arrangement, he balanced himself on one hand, and reached his geological hammer with the other. He ultimately dug this in, and made of it a prop by which he withdrew himself from a very nasty situation. Then he took the photograph, and thereupon continued his way.

Meanwhile I was wandering calmly over the lower slopes of Mount Marmier, seeking a good plane-table station, and ignorant of the adventure in progress below. Rounding a corner, the Grit Ridge Glacier came in sight, and beyond it

Grit Ridge itself, which belongs to the Lusitania *massif*, and is, in fact, parallel to and south of the ridge of that mountain, whose slope falls northward to Sassen Bay. Another

GARWOOD SLIPPING INTO THE TORRENT.

glacier, filling the hollow between these two ridges, and whose main torrent descends direct to Sassen Bay, close by Starvation Bluff, sends a snout over the depression between

Grit Ridge and Mount Marmier, and the stream from this cuts a deep gorge in the left marginal moraines and boulder-clay piles of the Grit Ridge Glacier. My route interposed this gorge between me and the direct ascent to Grit Ridge. It was a gorge with precipitous sides of mixed composition, sometimes cut down into the paper-shales below, sometimes blocked by great boulders of hard rock from Grit Ridge, sometimes falling by steps, over which the water plunged in wild leaps. I made many ineffectual attempts to cross it. Once when I was cutting steps down the steep side wall of compact débris the whole thing gave way, and down I went, only just arresting my fall at the top of one of the cascades. I ultimately crossed by a rotten ice-bridge. Garwood joined me on the glacier, and we went forward together. We concluded that solitary rambling in such regions is unwise. Sane persons do not ramble alone in the upper regions of the Alps, but we had not yet come to realise that wherever ice reigns, though it be but a few feet above sea-level, precautions should be taken which are beyond the resources of a single individual. This day we had come out, as usual, without a rope. As will be seen, we were destined to repent the omission.

The surface of this glacier, like all others we had thus far seen, was entirely and deeply covered with winter snow. It became a question of wading, not of walking. Reindeer tracks abounded, ascending even to the névés[1] and crossing high ridges; they were all fresh, doubtless indicating that the warm weather was leading the beasts up to cooler feeding-places. At all events, after the heat came, we no longer saw any reindeer about camp, whereas before they had been grazing around on all sides—a fact which, as will be

[1] I use the ordinary word névé to denote the upper basin of a glacier. There are, however, no true névés in Spitsbergen. All the snow that falls in a season is turned into ice before the next.

seen, was full of misfortune for us. Fox tracks were likewise numerous and recent. Brer Fox in these parts seems to be a "monstrous soon beast," leaving traces everywhere, but seldom visible in his own proper person. Thus far he had preceded us up every hill we climbed. He traversed Fox Peak. He was not only up Sticky Keep, but he went out of his way to scramble to the edge of the overhanging cornice at its highest point, apparently to obtain an uninterrupted view all round. We caught a glimpse of him, watching us at the pitching of Cairn Camp. This day we saw him again, but very far off, alternately stopping to look at us and then cantering away over the snow with the jauntiest gait imaginable.

The ascent to the lowest hill-top of Grit Ridge was dull and laborious, up a snow gully, the rotten shale-ridge by it, and the rottener slopes above, to the snow crest. Mount Lusitania rose opposite, beyond a large névé basin. It hid much of the desired view, and so became an enviable point of vantage; moreover, I had sentimental reasons for wishing to make the ascent. In the year 1894 the Orient Company's steamship *Lusitania* took a party for a cruise in Spitsbergen waters. After reaching latitude 80° 30′ north, they spent three days in Sassen Bay, during which a passenger, Mr. Victor H. Gatty, climbed and named this peak. He wrote an account of it and sent it to me, as Editor of the *Alpine Journal*, for publication.[1] I was thus led to look up the history of mountain exploration in Spitsbergen, and this was how my attention came to be directed to the region. Mount Lusitania was, in fact, the cause of our presence in its neighbourhood.

The day was now superb, with clear distant views, and such depth of blue even in the nearest shadows upon the

[1] *Alpine Journal*, vol. xvii. p. 309.

snow that they seemed like sky-carpets cast upon the earth. We decided to attempt the ascent. Lightly we set forth over the wide snowfield, imagining that it would resemble the others we had crossed. At first all went well; the snow was pretty good, and we sank only calf-deep into it. After half-an-hour we began to go in knee-deep. Then followed an area of mere flour and pie-crust—a surface too weak to bear our weight, but strong enough to resist the forward pressure of the knee. Advance became absurdly slow, and half the passage was not accomplished. Sometimes the ice-axe, plunged up to its head, reached the hard ice below, but often soundings failed to reveal any firm bottom. We erroneously supposed that this was because the surface snow was deeper than the length of the axe. The winter snow was still present in overwhelming quantity. If the snowfield we expected to cross to Agardh Bay were to prove in this condition, a week would be needed to traverse a few miles of it. The reflection did not add to our happiness. At last, when I sank in deeper than usual, an ominous tinkle was heard below. It was caused by icicles falling into a crevasse. This was more than we bargained for. We had not brought a rope. Careful inspection now revealed a maze of crevasses ahead, deeply snow-buried, whilst away to the right were the tops of a quantity of seracs.

We did not yet realise that there were crevasses all round us, and that we had been walking over a whole series of them for a couple of hours. Every time that the axe failed to touch bottom it was because we were over a crevasse, and not because the winter snow was too deep. This I discovered many days later, when looking down upon the glacier from the top of Mount Lusitania. All the fresh snow was then melted away from this area, and the network of crevasses was revealed in naked complexity. The schrunds were wider than the ice-walls dividing them, and I shuddered to think that we had actually walked, unroped, across so dangerous a place.

As it was, however, we knew enough to perceive the necessity for care. Advance became cautious as well as slow. Every step was probed in advance. Another crevasse was revealed, then a third, the farther side of which could not be felt. We tracked along it, but could not cross. An

PROBING FOR HIDDEN CREVASSES.

hour passed, and we were no nearer Mount Lusitania. The crevasse swept round the whole width of the glacier, and cut us off. There was no wise alternative but to go back. It was taken with regret. We returned in our steps past what we knew to be a crevassed area, then bent off to the right and climbed a higher point on Grit Ridge. The snow remained wet and evil. Its surface glittered with the mere tinsel of water-drops, not with such sparkling ice-brilliants as adorned the snows of Fox Peak.

Sometime after midnight the top of our second-choice

peak (2130 feet) was gained (July 7), a good summit of large firm rocks of a kind new to us. They formed comfortable and dry seats, where, in shelter from the light cool air, we could bask in the warm sunshine, and, removing wet boots and stockings, could let our cold feet feel the genial warmth. Here lunch was consumed, and two enjoyable hours were passed, valuable also for surveying purposes. A few clouds alternately assembled and dispersed, but they hid no portion of the glorious panorama, which included, from a new standpoint, only peaks, glaciers, and valleys already known to us. The snowy region at the head of the Rabot Glacier, in the remote distance, was better seen than ever before, and we thought we could distinguish Mounts Teist and Krogh and other snowy domes, whose bases are washed by the waters of Wybe Jans Water.

The descent was without incident. We kept together, and found a better ford over the torrent. Camp was reached about seven A.M. in the waxing daylight. Our last looks whilst preparing supper were towards Temple Mountain, whose rock face was being washed by light from the east, which so outlined its many ribs with bluest shadows as to render it the very semblance of a vast and splendid columnar edifice, rising upon a boldly sloping plinth from the blue surface of the bay. The gently rounded snow dome above completed an architectural suggestion, which genius, with opportunity, might work out into a monument of transcendent magnificence.

After some hours of tranquil slumber we awoke to find nothing changed save the direction of the wind, the incidence of the shadows, and the volume of our supplies. Glaucous gulls had been busy with our reindeer carcass.

"Crammed and gorged, nigh burst with sucked and glutted offal,"

they jeered at us as we came forth. They had dragged

the reindeer's skin twenty yards away down a snow-slope into the brook, and, with incomparable impudence, left it exactly where they found and carried off the joints. The snow was covered with their footprints, and with reindeer hair along the track made by the skin. Our food supplies being thus

MOUNT MARMIER AND THE COLORADO PLATEAU SEEN ACROSS THE SASSENDAL FROM MOUNT LUSITANIA.

reduced, the first business of the day was to go up the valley and fetch the reindeer shot by Pedersen, and, as we supposed, concealed. Coming in sight of the place, we saw afar off a cloudy white patch upon the ground with a glaucous gull in the midst. The patch was reindeer hair, strewn about. Only the skeleton and skin of the carcass remained. The rest had been eaten. We searched in all directions for the beasts which had been so plentiful a few days before. They were no longer in the neighbourhood; warm weather, I suppose, lured them aloft. The return to camp was rather less light-hearted than the departure had been, but we were presently cheered by the sound of a human voice, and peeping forth saw a man approaching over the plain, driving two ponies, and, by some freak of light, looking, as he drove, far "larger than human" on the sunny plain. It proved to be Trevor-Battye come from Sassen Bay.

He had landed, at ten in the morning, at Starvation Bluff. It was eleven o'clock at night when he came into camp, fairly tired out. On landing he met Pedersen with the ponies, and heard a wonderful tale of our miseries up country, and how we were on the point of dying of starvation—all absolute moonshine, but Pedersen was a born romancer. Sending the others back to Advent Point with the boat to fetch the sledges, and loading up the ponies with packs, Trevor-Battye hastened, as he supposed, to our relief. His march was no light matter. There were no pack-saddles, and the ponies were not used to carrying loads. They were frightened by the glitter of tins on their backs and the tinkling of spirits in them. The loads came loose, and things tumbled off and had to be cached here and there along the way. At last Bergen set to work kicking and bucking. His load shifted round under his belly, and he seemed to have become fairly frantic; finally Trevor-Battye cut the lashings in despair, and permitted the burdens to fall on the ground, noting the position of the place, as well as he could in the dense fog that then prevailed. Unfortunately Pedersen, who did the lading, had used the sledge-harness and our climbing-rope to fasten the loads. These were now cut into small pieces—a misfortune that in coming weeks caused us infinite annoyance. When Trevor-Battye at length reached Waterfall Camp he was the only burden-bearer, the ponies following him unloaded. With them came the sea-fog, burying all the valley out of sight. Garwood again went forth after reindeer, but it was a hopeless quest in such atmosphere. He returned empty-handed.

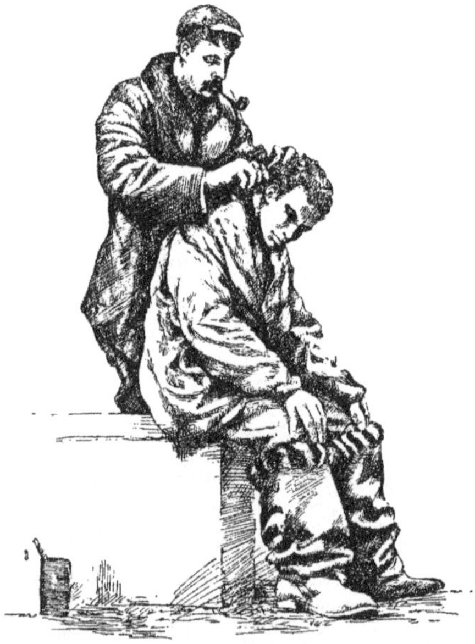

HAIRCUTTING.

LOOKING SOUTH-EAST FROM THE TRIDENT.

CHAPTER X

THE TRIDENT

THE rare spell of fine weather was in fact at an end, and the mist when it rose formed into higher clouds which hid sun and peaks from valley floors. This mattered the less, as we had work in hand which seemed to require no distant prospect. Camp had to be moved across the Esker Stream, and the things Trevor-Battye had been forced to drop on his upward way must be fetched in.

At eight P.M. (July 8) we settled down to our respective tasks, after a lengthy argument as to whether it was really eight P.M. or eight A.M., and whether it was to-day, yesterday, or to-morrow, so confusing does the lack of night become, especially when the sun goes behind a roof of cloud, and one cannot reckon, after the fashion of sailors in these parts, by the point of the compass at which it stands. Garwood and Trevor-Battye took the ponies and went down the valley, whilst I struck camp. They were lost in the mist a few yards away, and I was alone with the glaucous gulls. Pulling down and packing tents and baggage is slow and rather tough work for a single pair of hands. As each bundle was completed I carried it down to the foot of the hill, across a brook, and deposited it at the margin of the ford about three furlongs away. There were infinite small tins and bags to be bundled up somehow; having travelled hither on sledges, there were no sacks to hold them. After an hour or so the mist rose about a hundred feet off the ground, revealing the solemn brown and purple plain with leaden rivers flowing through it—a flat of apparently limitless extent, for all its margins ultimately faded into fog.

Eight heavy loads I portered, one by one, down the boggy hill-side. The first was something carried away from camp out into the wide world; but, as each load followed and fewer remained behind, as each tent was in turn emptied, and the canteen and store-tins disappeared, the sense of home was taken from the old place, and gradually transferred to the pile of baggage by the ford. Strange, how keen in the wilderness becomes the sentiment attaching to "one's things," the visible and transient connection that for the time links one with a particular spot and distinguishes it from all others! Their presence anywhere invests the place with a kind of consecration, as of the Aryans' sacred hearth. Remove them elsewhere, and the spot they quit reverts at once to its former aloofness. The stones on the ground, for a day known so well, give up their individuality, and become mere common fragments of the broken hill-side, not different from millions more about them. The camp-knoll melts into the landscape, and is unrecognised a mile away.

By midnight my work was done; all the baggage was grouped by the river-side, except a pile of geological specimens packed in tins, in charge of which, and of the reindeer skeleton, was a solitary skua, who croaked some kind of farewell at me as I finally turned on my heel. Five minutes later a crowd of gulls were in occupation of the abandoned site. The river, where we had to ford it, was swollen to over a hundred yards in width, with low stone-beds emerging from the water here and there. By selecting a zigzag course, shallow places could be found, so that the ford was nowhere deeper than a few inches above the knee. The water was ice-cold and the current rapid. I sat down tired on the bank and thought the prospect dismal. Long it seemed before the others returned; there was no sign of them over the dismal flat or along the sloping foot of Sticky Keep. Distant lumps of moss on the sky-line against the mist sometimes mimicked the forms of their heads. No breeze stirred, nothing moved save the muddy torrent. The only sounds were its treble babblings, and the faint bass note of the hidden waterfall; rarely a bird's cry broke the silence. A flock of eider-ducks flew by, but never a goose. The bereaved birds had gone away. A garden of little white flowers blossomed about me in the dry mud. There was no novelty amongst them. I sat on, smoking, and thinking of many things, as the stream flowed by, but with one eye fixed on any hill-slope that might be clear, in hopes to see a reindeer. None appeared; only grey erratic blocks or dusty snow-patches mimicking their form. At last the absentees emerged from a distant island of fog. The cooking-pot was set boiling, and, when they arrived, hot soup consoled them for the streams they had forded, and the fog that prevented their finding one of the piles of things, and thus deprived their labour of half its reward, and incidentally wasted another day of my time.

WADING A TORRENT.

About two A.M. (July 9), we shouldered heavy packs and entered the water with Bergen well laden. The passage was safely accomplished, the only misfortune being that on the far side the pony got rather badly bogged. A good site was found for the new camp (97 feet), and I remained to pitch it, whilst Garwood led back Bergen for another load. My work concluded, and the others not arriving, I went back to see the cause of their delay, and was exactly in time for the show. Trevor-Battye and Garwood were leading the ponies down the opposite bank, Spits with two evenly-balanced packs, Bergen (who before crossed so calmly) with a miscellaneous assortment of goods elaborately attached. All went well about half-way; then Bergen took fright and began bolting and bucking. He dragged himself loose from Garwood, and began pirouetting around, with his hind-legs or all his legs in the air, making bolts hither and thither, pausing for another series of bucks, and bolting again. At last his load flew in all directions, the whole of it fortunately landing on an island, where my despatch box burst open. Diaries, note-books, envelopes, ink-pots, aneroids, thermometers, boxes of photograph-films, and what not strewed the foul ground. By a miracle no important thing was lost. Freed from its load, the pony made off up stream through deep water, and landed at the mouth of Waterfall Gorge, whilst Spitz stood neighing, and the others split their sides with hysterical laughter. It was a comic scene, not unmixed with a tragic element, for it meant much tedious work to be done. I hurried after the truant, who fortunately became bogged and was easily overtaken and brought back, but all attempts to reload him only brought on convulsions again, and we had to do his work. The scattered things were gathered together. The contents of the despatch box were laboriously recovered. The case itself, an old cartridge-box, still grimy with Himalayan and other mountain dust, rubbed into its canvas cover at various times and in many places, was now coated with Arctic ooze.

When Bucking-horse Camp was pitched and cooking well in hand, after three hours' toil in and about the ford, Garwood, who had done the lion's share, told us of a certain miserable day spent geologising in the North of England. He was awaiting a train in Appleby refreshment room, when a dripping drover entered. "Give me some hot water, Miss," he said to the barmaid, "and some sugar and plenty of brandy. I'm sick of this blooming world." We were all sick of this blooming world that night, but there was no consoling bar at hand, and water took long to boil in the rising wind. Ultimately all was comfortably arranged. We retired to rest with the dark outlines of various drying garments visible through the semi-transparent roof of our tent, and looking like so many misshapen torsoes and amputated limbs.

Cold blew the wind through the hours of rest, and miserable was the chill and cloudy weather to which we awoke. I had journals to write and a tent to mend. The others went forth to look for birds, fossils, and especially for a reindeer. Writing in camp was frigid work. I retreated into a sleeping-bag for warmth. Time passed, and work was done, but without incident or satisfaction. The darning of stockings is doubtless an occupation not without charm, but to lie on your face in a freezing wind, and drive a packing-needle through sail-cloth and rubber sheeting (when it does not go into your hand), is a wholly disagreeable employment. It suggested an inquiry to which no solution was apparent: Why is it so much easier to drive the blunt head of the needle firmly into the hard object you use in place of thimble, than to force the sharp point through the relatively soft material it is intended to pierce? The fact is indisputable, the cause obscure. Another curious fact was brought to my notice by the day's experience. It would have been supposed, considering the number of generations there have been of sewing humanity, that sewing tackle would be as nigh perfect as possible. Yet apparently it has been left to me to discover that the head of an ice-axe is far superior to any thimble for the purposes to which thimbles are devoted. Sewing at the luckiest is a dangerous employment, and should be so scheduled. Every hus'if ought to contain sticking-plaister. Mine does, and this day it was largely employed, and very useful.

Before midnight the sun shone brightly, but there was no warmth in his beams. At the same altitude above the horizon in India his rays would be hot enough to give a sunstroke. His spare efficiency might be better distributed. After midnight I was free to wander forth, and use the renewed clearness for survey purposes. My way was up the main valley across the lower slopes. It was delightful to find a mile of fairly hard ground on which one could stride freely, and feel the joy of visible advance, instead of creeping painfully from step to step. The Trident's jutting bluff aloft showed some boldness of front, and thrust steep-seeming crags into the sky. The cool temperature and breezy air, which made work in camp so unpleasant, were exactly right for exercise; blood coursed bravely through the veins, and activity brought joy to the mind as well as comfort to the body. A mile or so along, near the mouth of the next small side valley, Garwood came up with pockets full of fossils, gathered in the second small side valley, where he and Trevor-Battye had seen two shy reindeer, which they failed to secure. Trevor-Battye went on to Turnback Valley, and came into camp later, when supper was cooking, and the tents were set in order for sleep.

His observations led him to conclude that, in the matter of birds, there was no new work to be done in the interior, whilst the flora of the inland valleys and hills is identical with that of the coast. Nothing, therefore, attracted him to remain with us. He accordingly determined to take the ponies down to meet Gregory at Sassen Bay, and to utilise the boat during our absence for the purpose of visiting Dickson Bay, or such other points on the shore of Ice Fjord as might afford opportunities for his special studies. He undertook either himself to meet us at Sassen Bay in ten days' time, or to send a boat to await us, if he could hire one from any sloop or at Advent Point.

Accordingly, in the afternoon he and I set forth with the beasts, whilst Garwood went to climb the North Prong (2172 feet) of the Trident, whence he returned in due time with many photographs and some useful topographical and geological observations. We, meanwhile, waded the ford yet again, and plodded down the valley, going first over the Esker's spongy raised delta, and then along the bog-slopes, always looking out for the abandoned bundles. At last we found them and halted to divide the spoil, part of which I loaded into my rücksack, part left for future use on our return, whilst the remainder, consisting of clothes and the like, was packed with much careful balance and solicitude, on the backs of the now unresisting beasts. After starting Battye on his farther way, I left him, and retrod the track, bearing a precious load of salt, onions, two bottles of beer, oatmeal, Irish stew, and such like, the burden being made light and the way short by foretasting the pride wherewith I would display such treasures to the admiring Garwood. The deserted camp, with one tent prone, looked forlorn indeed; there was a note from Garwood detailing his plans, and the sun was shining brightly, but never a pony neighed nor a living thing stirred in the large solitude, save when some fulmar petrel came by—one of the long procession now continually passing up the valley on the way to the east coast. Sitting silent in the tent, one could hear from time to time the whirr of their strong wings as they passed close overhead.

I brewed a cup of tea—no brief task when water has to be fetched from a snow-hole, into which it but slowly trickles—the lamp filled with spirit, cup cleaned, boiling brought about with the normal deliberation of the watched pot, the compound strained, milk and sugar dislodged from their hiding-places, and all by one pair of hands, within the confined limits of our shoulder-high tent. In half-an-hour the work was done. Then came the preparation of a mighty supper. Into the pot went the shredded fragments of two onions, a handful of dried vegetables, odds and ends of arrowroot and oatmeal, a lump of bovril, a seasoning of Worcester sauce (all spoils from Trevor-Battye), and ultimately, when Garwood's jodel was heard on the slopes, half a tin of Irish stew. Ye gods! what a jorum it was! and how it and the fried slices of rich plum-pudding that followed suited the complaint of two hungry mortals, whose food for many days had been stringy reindeer or concentrated rations! The charmed, unsetting sun looked down upon us, warmed the soles of our feet, and dried our garments. The wind slept. It was an hour of peace and perfect charm—light, colour, air, scenery, all fair and pleasant to every sense, rare combination, nor in Spitsbergen only.

Such times do not long endure. The spell was broken by a puff of wind that rattled the tent, harbinger of a gale soon to follow. All our Spitsbergen camps were of necessity in exposed positions. We found no sheltering rocks, whilst all hollow places were boggy. Dry ground was only on protruding knolls, bare to every wind that chose to blow. The gale that opened fire on us after midnight boomed against our canvas walls and roof, and made the stretched ropes sing like the strings of an Æolian harp. At first we doubted whether the ropes would hold, but confidence grew with experience. All the while the sun was hot, far too hot for comfort in fur sleeping-bags. Light, heat, and noise kept slumber long away, and made our waking correspondingly late.

Next day Gregory was expected from Advent Bay, but we doubted his being able to bring round the boat in the gale that continued to blow. So, fired by Garwood's description of the interest of the view from the Trident's crest, I set off with him to climb the central point of the three ridges, whose precipitous north front falls to the Sassendal. We mounted along the side of the valley, turned up a gorge filled by a gentle snow-couloir, between the west and central prongs of the Trident, and climbed to the col at its head, whence an easy ridge led to the top (1990 feet). For a time we left the gale behind, and could hear it tearing over the crags below, but when the plane-table was set up, the temptation was too strong to be resisted, and up came the wind to spoil our pleasure. Not that there was much fun this day for Garwood; he was suffering tortures from neuralgia, which he bore with heroic fortitude.

Here at last we had a thoroughly intelligible and wide-extending view, full of most valuable information, which cleared up all manner of topographical puzzles. Bunting Bluff and the other bluffs west of Advent Vale were all in sight with their upper prolongations and high snows. We could see almost down to Bolter Camp. Fox Peak was hidden, but this was the only loss. More important than these points, the correct position of which on my map I was almost as much surprised as delighted to verify, was the area to the south and east that lay open below. With our backs upon the Sassendal, we had before us a long slope and an undulating country below it, the form of whose surface was determined when its drainage ran down Advent Vale. Now the Esker and Turnback rivers have cut back and robbed most of these waters. This lower, undulating country is bounded by the glaciers and snowy hills near the Baldhead, which spread away eastward till they merge in a high glacier region, practically an ice-sheet, apparently stretching as far as the east coast. This ice-sheet sends several glacier tongues down into Turnback Valley and its branches. One of these, of gentle slope, seemed so easy of access at the foot that we thought it might prove the best avenue of approach to Agardh Bay. As yet we did not know of the existence of Fulmar Valley, and were still labouring under the belief that an ice-sheet must be crossed before we could arrive at the east coast. The prospect was a noble one in all directions, but the freezing wind robbed it of charm and made study of it painful. An hour's work was all I could endure.

On first reaching the top we noticed a strange white bed of cloud, all isolated, covering a level place in the midst of the névé below the snow-pass to Agardh Bay. It was a bright round silver cloud, resting like a bubble upon the snow. When it lifted two giants were revealed, a man and a graceful and beautifully-draped woman in floating lilac garments. They were both about two hundred feet high, and there were giant children with them, also dressed in lilac drapery. They seemed to dance upon, or rather to float over, a wonderful dark-blue carpet, laid in the midst of the snow-field. They moved very slowly, but with indescribable grace, and the woman sometimes hovered over the man with her veil falling in a glorious curve over her extended arm. As we looked, they were gone, and in their place, in the midst of the blue carpet, stood a magnificent tree, also lilac in colour, with a great trunk and a huge extending top, in form like one of Turner's pines. It also vanished, and other strange forms appeared and changed, coming and going with mysterious rapidity. They were all varieties of a single cloud, kept in one place by the eddying of the air and casting the dark shadow, so much darker than itself, upon the snow, which it almost touched with the point of its lower extremity. It was a weird phenomenon and of rare beauty, but even it failed to keep us on the freezing peak. During the descent a white falcon was seen, soaring far aloft, and a reindeer in the Sassendal, which made off the moment he saw us.

Camp, when we returned to it, was still deserted, nor did the closest scrutiny reveal signs of Gregory's approach. "He cometh not," we said. The wind still raged and boomed, blowing dust into the tent with horrid persistence, but the canvas bravely resisted, and formed a delightful shelter when the doors were closed. Later on the wind dropped, but then all the sky clouded over and rain began to fall (July 12). Its pattering on the roof was a sound of peace that lulled us pleasantly to sleep.

Meanwhile Gregory was toiling up with the ponies and sledges from Starvation Bluff in Sassen Bay. He had to ford swollen rivers and to face the driving rain. It must have been a miserable march. Williamson drove the second pony. The two arrived when we were in the midst of our deepest slumber. Then arose much cooking and setting of things to rights, and not a little exchange of news, for a steamer had come into Advent Bay before Gregory left and brought mails from home—weekly editions of the *Times*, illustrated papers, and letters. Sleep was banished while other interests reigned, and Gregory told the tale of his wanderings, much as it is recounted by him in the following letter :—

GREGORY STARTING FOR ADVENT BAY.

"YAMBUYA CAMP, *July* 9, 1896.

"I ended my last letter thinking that three hours later our boat would be taking it back from Sassen Bay, and that I should be on my way inland with two strong sledges, which would enable us to cross the glaciers to the East Coast. But the chances for delay in this country are infinite. It had been arranged that our boat should be at a given station in Sassen Bay on Friday afternoon, and I accordingly left Waterfall Camp on Saturday afternoon, feeling no doubt that the boat had already arrived. When, therefore, while wading a river, my pockets were filled with water and my biscuits turned into sop, I did not mind; and only laughed when the enterprising pony Spits seized the opportunity of their being spread out on the sledge-end to dry, to devour the lot. I was a little more concerned when, on arriving at the trysting-place, I found no sign of the boat. There was much loose drift ice in the mouth of the bay, but Pedersen, in his broken fo'castle English, declared that no weather could be more favourable for the run up the bay. I therefore decided to stay on the shore, and wait at least twenty-four hours for the boat. I had no sleeping-bag, but as the night was fine that mattered little. I was more disconcerted by the fact that our only food consisted of three odd fragments of Emergency Food and six small sticks of chocolate.

"This supply I resolved to keep till the next afternoon, when I intended to share it with Pedersen, and, on the strength of it, march back to Waterfall Camp. I tried to shoot some birds, but after knocking the feathers out of an eider drake at a distance twenty yards too great for a kill, I had no other chance of a shot. There were, it is true, a couple of snow-buntings resting on the cliff, but they sang to me so merrily that I could not kill them.

"For once in a way the sun shone gloriously, and I had also the good luck of a magnificent view. At the head of the bay was the broad Post Glacier discharging icebergs into the fjord; on the shore opposite me rose the vertical stratified cliffs of Temple Mountain. The enjoyment of the scenery helped me to forget that it was cold, and that the water of two rivers and a dozen brooks was gradually evaporating from my clothes. At midnight I wrapped myself in a sledge cover and tried to sleep. But the air, though delightfully crisp and bracing when I was moving, was so chilled by floe and glacier, that I was soon far too cold to sleep. When my teeth commenced to chatter, I thought it time to go for a walk, as I dreaded another bout of fever. Just as I was preparing to start, at two A.M., I faintly heard three shots fired in rapid succession. I woke Pedersen, who, having his sleeping-bag, had been sound asleep since nine o'clock. We both heard a fourth shot, to which we replied by one as a signal, and then I sent Pedersen in search. The suggestion of food was enough to start the hungry hunter in all haste. By this stroke of luck I felt sure of news and food in the future, as well as of the immediate benefit of a warm sleeping-bag. Inside it I was more comfortable, but was too hungry to sleep. I turned out all my pockets in the vain hopes of finding some stray scrap of something eatable, and then counted again the sticks of chocolate. I dared not eat any of these, as we might have to rely on them to act as a stimulus for the wade back through the bogs of Sassen Valley. At ten o'clock Pedersen came back. Instead of meeting our boat party, he had found that of Baron de Geer, who had generously sent me some reindeer meat, cheese and biscuits. There was ample supply for one man for two days, so I resolved to leave Pedersen here to watch for the boat party, and myself return to Waterfall Camp, and thence march overland to Yambuya. Pedersen and I therefore ate as small a breakfast as possible of biscuits and cheese. I waited till the afternoon to see if the boat came in or the ice packed with the change of tide. At four P.M., leaving a note for the boat party, I started back for Waterfall Camp, arriving there ravenously hungry at 7.30. I had one mighty meal to clear off arrears, a long sleep, and then another big feed to lay in a reserve for the future. At 5.30 P.M. on the Monday, I was ready for what Conway called the colossal march to Advent Bay. This had taken five days to do on the way over, but then it was unknown ground, and we had the sledges. It had now to be done in one, partly in hope of reaching Advent Bay in time to catch the boat, partly because I could not carry much with me, and sleeping in the open air without a blanket does not agree with me.

"I left Waterfall Camp at 5.30, and started up the Esker Valley for Brent Pass. I carried with me a little Emergency Food and fried reindeer meat, but trusted mainly to Root's Cuca Chocolate, which, as it does not make one thirsty, I find the best thing for forced marching.

"A sharp walk of two and a half hours brought me to Brent Pass, where I rested for half-an-hour to take food, sketch the hills to the south, and write out notes on the geological facts seen during the ascent. I had intended to shorten the journey by keeping along the right bank of the river, and fording this near Advent Bay. But one glance at its channel showed me that I must cross at once; for the powerful sun of the previous days had melted so much snow, that the river volume was many times greater than when we had crossed before. Fortunately the river channel here was a broad stony plain, more than a mile in width, and over it the river ran in many channels. I forded fifty-two before reaching the ice-foot on the left bank; many of the channels were small, but several were fifty feet wide, and one came up to my waist. Fortunately after this a narrow band of ice ran along the bank of the river, and over this I was able to make rapid progress. But as the river channel narrowed, the stream occupied it all, except for a few shoals in the middle, and I was driven on to the rough, boggy rock-strewn floor of the valley. Numerous deep gullies were cut through this to the level of the river; some of these still contained much snow, and the snow bridges were strong enough to let me cross the rivers without the nuisance of a wade. Farther on the walking was better, for the ground was dry and stony, and by a spurt I reached the old terminal moraine at 11.30. I had taken some biscuits from our food cache at Sunshine Camp, and I rested on the moraine till midnight, nibbling kola-biscuits and chocolate, and sketching. The geology here gave me the clue to the structure of the whole valley. So I went on again refreshed by my half-hour's rest. An hour's walk took me to a point level with Cairn Camp, where I found myself on an old marine beach, so that the fjord once ran up to this point, and probably the old glacier then entered the sea here.

"The next point of interest was finding a Turnstone (*Strepsilas interpres*), which has been twice before seen in Spitsbergen (by Dr. Malmgren and Professor A. Newton), but so far as I know has never yet been collected here. I had no gun, and so could not secure the specimen, which was irritating. I had now seen twenty-four of the twenty-seven reliably recorded species of birds; all but the Turnstone were of little interest. Finding this bird so early in the year shows that it probably breeds in Spitsbergen.

"After this I had six rivers to wade, one from each of the valleys that come down from the plateau between Ice Fjord and Bell Sound. The smallest of the six rivers had twenty-two channels, excluding those which I could jump. Twice I was nearly knocked over by the force of the current, for the streams were powerful, and I was getting tired. So far I had taken very little food, and only rested twice, but now it became necessary to take a rest of ten minutes once an hour, and eat some chocolate at every rest. But Advent Bay lay below me, and the tents of Yambuya were in sight. At 5.45 A.M. my hulloo woke Studley, and frightened away an Arctic fox, which was stealing reindeer meat from the rack beside the camp. Studley at once made me some tea, and while sipping this he told me news of the rearguard, and I told him the story of our traverse to the Sassen Valley, and of my walk back. This had taken me twelve and a half hours in all, including one hour and forty minutes' rest. The weather had been so glorious, the scenery so fine and the geology so interesting, now that the mists had cleared away and I could get more than glimpses of the country, that the walk had been delightful. The only thing that marred its pleasure was to find at the end, that, instead of being able to return the same evening to Sassen Bay, I was doomed to a few days at Advent Bay. And if there is one thing that does not suit me, it is work in rear-guard camp."

ADVENT POINT CAMP.

CHAPTER XI

FULMAR VALLEY

WE were now supplied with strong sledges and needful food. The onward way lay open ; but both Gregory and the ponies needed rest, so the start had to be postponed twenty-four hours. Garwood and I used the interval to revisit Brent Pass. The walk was dull. We followed the right bank of the Esker Valley and had to wade many knee-deep streams. A marked change had come over the place since our former visit. Snow-beds and snow-bogs had entirely disappeared. The land was drier and harder. A distance that had been covered in the most laborious march we made, which utterly exhausted the ponies and wearied us, now only required two hours or so of steady plodding. Under any circumstances we could cover in a day (if unloaded, and with no survey or geological work to be done) three times the amount of ground that ponies can drag sledges over. Hence the apparent slowness of our progress. Clearly by the middle of July conditions are much improved for travel in the interior of Spitsbergen.

The day was cloudy, with heavy and rather high clouds, which cast a pall of gloom on the dreary landscape. Yet there were fine effects to reward an observant eye. A soft grey cloud lying on the Trident dyed its precipice-headed bluffs with a purple, so rich and deep, that it seemed more than any mere light and shadow could produce. Distances seen beneath the cloudy roof showed continual and most beautiful changes, as, for instance, down towards Sassen

Bay, where strange lights played on the water, and gleamed up under the mist, or at the head of the Sassendal, where Rabot Glacier glimmered like oxidised silver set in ridges of bronze. The brown plain between was always mysterious, with quick alterations in apparent size, whilst now and again eddying cloud-columns dipped down on to it, dissolved, and disappeared.

At 7.45 P.M. (July 13) we finally quitted Bucking-horse Camp, leaving a pile of provisions, geological specimens, tins of spirit, and other supplies to await the return. Two skuas remained in charge, greedily picking the backbone of a reindeer. They are very wide-awake, these skuas, and not in the least shy. A number of fulmar petrels came along, bending in the direction we were taking. Their swift, easy motion and strong confidence of wing are always a delight to behold. Overhead the cloud-roof was denser than ever, but it was poised high above the tops of all the peaks, and cast on the whole landscape that dark rich pall of colour which so dignifies these valleys, diminishing indeed their appearance of breadth, and shortening distances, but at the same time raising the bald hills into the likeness of mighty mountains.

Descending the last slope below camp, we trailed along the plain some distance from the valley side, thus winning the best travelling ground we had yet seen in Spitsbergen, and gaining a view far back over Sassen Bay into the hills behind it. There a large valley, running inland, northwestward from Skans Bay, prolongs the Sassendal depression far off towards Dickson Bay, whilst the hills on either side of it imitate the forms of Temple Mountain on the right and Sticky Keep on the left. As usual I was soon alone in the bare valley, out of sight of the others. It seemed barer and more uncanny than ever this gloomy evening. A strange cry came from over the river, like the cry of an abandoned child.

Then there was a sound in the air close above my head as of the rushing by of some great creature, some "ghost from an enchanter fleeing." A petrel had flashed past. All was still that the eye beheld. Hour followed after hour, and the view was ever the same. Across the valley the same long, low line of bare flat-topped hill; on this side always the staged fronts of the Trident, flat-bedded, each bluff repeating the forms of the bluff before, and striped with regular lines of exactly similar snow-couloirs—Peak Milne-Edwards in front coming so slowly nearer. At last, at midnight, the corner of the Turnback Valley was reached, and a gloomy prospect opened over a short wide inlet into the hills, with a broad flooded river flowing down the midst in many channels, and in places almost covering the half-mile of stony flat that intervenes between the gentle bog-slopes of either bank. Many small snowy side valleys were revealed, one of which we believed led to our pass. We afterwards discovered that it was not the one we had noticed from the Trident; but for the present we were deceived. The nearer we came to it the less we liked its appearance. Formerly it contained a glacier, which deposited a quantity of moraine all about its wide snout; this rough accumulation blocked the lowest half-mile of the hollow. Behind came the shrunken tongue of ice leading up to the great sheet. It was a miserable outlook. There appeared to be bogs to cross, the wide river to wade, a long stone-fan to ascend, and then the moraines to scramble over—a combination of all the nastiest things that can be put in the way of sledges.

Into the water we plunged. It was too swift and deep where I tried it, and carried me off my legs. A better place was found, and over we went from one shallow to another with broad deep channels between. In the deepest of these both sledges went gaily floating down stream, rolling over and over; whilst in the rapidest place, where the icy water rushed in waves up our legs, Spits took it into her head to stop and drink, and no struggles of Garwood could get her to stir. There were marks on the banks showing that the waters had recently, in the hot days, been at least a foot deeper, so we were to that extent lucky. The moraine when reached was less toilsome than we forebode. A dry and fairly smooth hollow was found leading into it, and there we camped (282 feet), for farther in was no grazing for the ponies. Moreover Spits had begun to go lame from rheumatism, and appealed for rest. She was now so much one of the family that she really spent most of her time in camp, with an eye constantly on the alert for biscuits. She would put her head in at the tent door, or rummage our things over with her nose, trying this and that. A bundle of tie-on labels attracted her special attention, and she never could quite make up her mind whether it was edible or no. On the march she co-operated with us rather than was driven, and had her own ideas about the way. If she was crossed, and led contrary to her judgment into a troublesome place, she would halt and look round reproachfully, as though to say, "I knew it." Her own opinion was by no means always right. When she made for some grey streak of grass, thinking it to betoken a dry place, and lo! it was bog, no one could more emphatically assert annoyance than she did. When hopelessly bogged she would not struggle, but, remaining quite still, would look round for help, and so wait till it came. Bergen at first had none of these arts and graces. He was a mere bundle of nerves, shy of everything and hard to tempt even with biscuit. He came to trust some of us a little by slow degrees, but the least unexpected tap or glint of metal would put him into a paroxysm of terror. On the other hand, he did not lose his head in the bogs, and pulled his load without fuss and with much energy if everything about him was normal.

When camp was pitched and the day's survey completed we had a great discussion about the route to be followed. We still did not realise that we had come to the wrong glacier, but we knew enough to argue that the pass, at the foot of which we were encamped, would not lead to the east coast at all, but only take us over into the valley (Fulmar Valley) on the other side of Peak Milne-Edwards. Pedersen, however, had related a tradition that this pass leads to Agardh Bay, but now we were come to it we more than ever doubted the accuracy of the information. Ponies could go no farther this way, and we must drag one sledge on ourselves. If with that labour we only reached the next valley we should be fooled, for the ponies (now we were over the river) could drag both sledges round as far as that by way of the Sassendal, and save us the trouble. The decision come to was to climb one of the neighbouring peaks next day, and have a good look round before actually plunging into the difficult area of snow.

Never had we camped where there was less grazing for the ponies. The moraine itself was apparently almost barren. The hill-side beyond it was plentifully covered with the vegetation that grows on wet bogs, but with this the ponies would have nothing to do; with fear and trembling, therefore, we were constrained to permit them to wander each with a stone hobbled at the end of a long rope to his fore-leg. Not that we were afraid of Spits making off on her own account. She was rather inclined to come into the tents and make herself at home, but she always followed Bergen, and his record was bad. He bolted the first day we landed, and he left us in a most informal manner at Bolter Camp. He had probably been ill treated by his late owner, and regarded the human species with distrust. Of late he had begun to be less hostile, and would even move a step to take a biscuit if quietly offered to him. The taming process went slowly forward, and we thought that after a feed of oats the two might be spared together to go cadging around for food on the hillside. They were accordingly led away and turned loose. What was our surprise, an hour after we had gone to sleep, to find them both come stumbling over the tent ropes, Bergen leading the way! They preferred our company and occasional plants on the moraine to anything the boggy hillside yielded. The morning found them close to camp, and almost warm in their greeting when the hour for harnessing came.

"The morning," I write, the result of habit; it was really the late afternoon, for we did not actually leave camp till 7.30 P.M., and then not for a peak, but to return down Turnback Valley to the Sassendal. The cause of this change of plan was, of course, the weather, which was so densely cloudy, that to ascend for a view was mere fatuity. Further reflection confirmed the opinion formed at supper, and we unanimously agreed that our right course was to continue the ascent of the Sassendal as long as possible, and to take the last considerable side valley that headed off in the Agardh Bay direction. In this the event proved our justification.

We vowed to make as long a march of it as possible, and to find our pass before camping, for we were all at last in as good training as a man can be in Spitsbergen, with rather indifferent and insufficient food, continual wettings, and a life in the open air indeed, but in an air that depresses rather than stimulates, and is wholly different from the bright quickening atmosphere elsewhere associated with snow-mountain regions. The ponies, too, were in better lasting condition, and Spits's lameness had almost gone. It soon became apparent that our wadings were not ended, indeed we had this day more streams to ford than ever. If they were not so wide as others we had crossed, some of them were quite as deep, so that all of us were wet above the knee ten minutes from the start, and remained wet the whole day, with puddles of water squdging and plopping inside the boots at every step. Besides streams there were infinite dry, or almost dry, gullies to be crossed, each from one to three yards deep, and with vertical sides, save at intervals requiring to be sought, involving the sledges in a zigzag route. Williamson, who is not over-burdened with brains, was inclined to take the sledge he drove straight across country, to its no small peril, for at one moment it would be standing on its snout, at another on its stern, whilst it frequently turned turtle. The iron hooping we had nailed underneath preserved it from a swift destruction. Garwood, who always acted as driver to Spits, conducted the rough sledge which we bought from the winterers for a few crowns. It was not much to look at, but it held together wonderfully. Its chief defect was a tendency to dig its nose into the ground, so that it often acted more as a plough than a slider. Much pulling and lifting thus befell, a work to which Gregory devoted the best part of the day.

On the whole, the ground traversed was the best we had yet passed, for the season was now visibly advancing; snow had almost left the valleys and melted out of the gullies, so that snow-bogs were become things of the past, and the soft area of soaking ground which we were wont to find at the foot of every snow-couloir was either greatly diminished or wholly dried. Instead of the snow-bogs were the gullies I have described, with their steep soft sides of powdered paper-shales; whilst the place of the larger flat snow-beds was marked by areas of soft sticky mud, seldom more than ankle-deep, and generally avoidable. All this mud, and, in fact, the ground everywhere, was cracked or cracking into roughly hexagonal lumps and bosses, either with or without vegetation growing in the cracks, according to situation and the nature of the material. Every earthy patch in Spitsbergen that we saw is characterised by this phenomenon.

Of course, as we had no rifle with us, the reindeer were more numerous and stupid than ever. They came near in sixes and fours, but there was never a fine head amongst them. They seemed more than usually muddle-pated and changeful of mind. On first beholding us they would approach, deliberate, and come to the conclusion, "These two-legged beasts are dangerous." With that they would turn tail and hurry away. In half a minute doubt would arise in their minds: "Perhaps they are not dangerous after all;" so back they would come to reconsider the problem, never adhering for a hundred yards' trot to any opinion or decision. Solitary bucks were the only exception to this rule, and occasionally a mother with her fawn.

As usual I went ahead with the plane-table, and this day kept the lead longer than usual, for few new points came in sight, after rounding into the Sassendal, till the time for quitting it. A wilder sky than of wont roofed the valley; dark clouds came sweeping over just above the summits, casting gloom upon the bare landscape, and every moment threatening to pour down a deluge of rain, which never actually fell. The march gave little incident and slight variety of scenery, for there was always on the far side of the valley the same changeless long rock-slope reaching in one smooth widely bending curve from far up the Rabot Glacier to Sassen Bay. On our side, blunt-ended bluffs followed one another, all built alike, with the same regular succession of strata emerging at unvarying distances above one another, and forming a series of shoulders, ledges, and knees. Yet as one looked down the Sassendal, this same row of bluffs produced a more striking and less monotonous effect than might be expected, thanks to the valley's wide curve and the variety thus introduced by perspective. The last view backwards, from the corner of Fulmar Valley, embracing the whole extent of the Sassendal, was dignified and large. At the corner an intrusive sheet of igneous rock, cutting a little crookedly through the beds along the east side of Fulmar Valley, introduced a slight novelty into the mountain architecture.

It was a relief to find a big river emerging from Fulmar Valley, for it proved that this was no mere short gorge leading steeply up to a high snow-pass or plateau. In fact, the Sassendal's river above the junction became relatively small, and flowed so smoothly that the opposite hill was reflected from its waters, a new thing to see in this island of torrents. It drains only the Rabot Glacier, a large one indeed, but glaciers do not here supply torrents so copious as those which at this time are filled by the meltings of the lower snows and the oozings of boggy hill-sides. Below the glacier's snout was still a huge ice-foot a mile or more wide and about three miles long, the most considerable example of frozen snow-bog we had yet beheld. A fairly wide rim of muddy rather than rocky moraine intervened between the clear ice of the glacier and this ice-foot. As the corner of the Fulmar Valley was approached the topography of the Rabot Glacier became better apparent, and two large tributaries were distinguishable coming in round corners from the north, where a great névé area must be hidden ; but I was in no condition of mind to heed closely what might be the topography thitherward, for the moments of turning the wide corner were moments of suspense.

The chief object of our journey was to investigate a complete belt of country across Spitsbergen from sea to sea. Unless we crossed the island ourselves, this could not be accomplished. If it were accomplished, we might consider ourselves successful. If it were not accomplished, any success we might attain could only be partial. This corner was the critical point. What kind of a valley should we find? Poniable or not poniable? Would there be a long glacier to surmount?—a high snow-field to cross? Some kind of a snow-pass we expected. Its position was the important consideration. Every mile it lay farther east the better, for we could push our base on, and thus diminish the length of the final rush. Of course if you can start a sledge on a smooth, gently sloping ice or snow sheet, and the surface is in such condition that the runners will glide over it, advance is easy. But our experience of Spitsbergen snow-fields was utterly discouraging. The snout of no inland glacier we had seen was at all easy of access. It would take two days' hard work merely to hoist a sledge and its load on to the foot of such glaciers, whilst a mile a day would be rapid progress over such snow-fields, for men and ponies could not have gone a yard.

Suspense and keen curiosity were the predominant partners in my mental condition as I hastened towards the corner. Flights of fulmar petrels kept passing along, and all turned up the valley. This, then, was their way to the east coast. At last the intervening angle was turned, and the desired view opened. It was a joyous surprise. The valley was wider than we had dared to hope—the widest tributary of the Sassendal—and it stretched away back to a distant turning in a true line for Agardh Bay. Fully half our remaining route was in sight, and no névé closed the prospect, whilst only the snouts of side glaciers protruded on the sight. The more of these the merrier, for the big torrent had to be accounted for, and if it came from the valley's side, it had not to be supplied at its head. Overjoyed, I turned back to meet my friends. They were far behind. First came a series of reindeer, preceding them like videttes, then, *longo intervallo*, the phlegmatic Williamson with Bergen and one sledge, both utterly bored, and taking a line like a Roman road, straight over every obstacle. At last the others came up, and we went on together. A new spirit entered into all; we vowed we would see the whereabouts of the pass before pitching camp.

Little had we thus far cared what the scenery of the Fulmar Valley might be; but as we advanced up it we had time for observation, for the way was long. Of all Spitsbergen's dreary dales, it must assuredly be the dreariest. It is the highway for every wind of heaven, so that snow never lies deep in it, and the ground is naked and less wet than usual. Sparse indeed is its vegetation, and late to put forth; the flowers were fully three weeks behind the Sassendal's, and many were only just blossoming which had already faded elsewhere. But this dryness and bareness of the land only made advance easier. Had it not been for the many side glaciers and the consequent number of streams to be waded, the walking would have been almost pleasant. As it was we pegged on doggedly, and even the ponies seemed to share our temper. Garwood larded the lean earth with puns more plentifully than usual, and no one complained even when, after catching and bottling some insect, he declared that we were the most agreeable party possible, because we put every living thing we met into good spirits.

On we went, gleefully watching the shrinkage of the river as stream after stream was crossed, and never a snowfield appeared at the end of the vista. At last some one was even bold enough to suggest that there might be no snow-pass to get over at all, and that we might take the ponies all the way to Agardh Bay. Every one objected that this was impossible for an infinity of reasons, because each hoped it might be true, and feared the disappointment of the rising hope. I alone was in pitiable plight, for the valley twisted all the reference points of my survey so swiftly out of sight that the plane-table had to be set up every half-mile in order not to lose connection with the country left behind. Thus I fell farther and farther to the rear. It was with no little satisfaction that at length I beheld the others halt, and saw in the far distance pitching of tents, and ponies turned to graze. Surely they cannot there be actually on the pass! Nor, in fact, were they; they had but reached the last visible grazing patch, a miserably barren ground, with low plants every here and there, a yard or two apart, in the narrow desert between the river and the débris slopes.

When camp was pitched and cooking in progress under Garwood's skilled management, Gregory went on round the corner, and returned with excellent tidings. There was moraine stuff from a large and perhaps final side glacier to be crossed, and that appeared to be the last impediment. There was no sign of snow-field to be passed, nothing that need involve the ponies being left behind. Thus encouraged, the day's concluding work of writing, map drawing, cataloguing, and the like went swiftly forward. It was 4.30 A.M. (July 15) when Delusion Camp (308 feet) was pitched, and at 11.30 the last closed his note-book, and turned over to woo the difficult god of sleep, with a milk-white cloud and a howling wind around the fluttering tents. The canvas pressing on the poles made them bend and quiver, and they in turn passed down their vibrations to the springy ground, which trembled beneath us, as though shaken by a series of small earthquake waves. The wind was a matter of no importance, but the fog was serious. It endured hour after hour, and was as dense as ever when we awoke, so that to go forward surveying was impossible.

Not till 10.45 P.M. were we able to set out, and then slapping our sides and blowing on our fingers, for the air was bitterly cold. The caravan plunged at once into a moraine area, the largest and most troublesome we had met;

AN EASY SPELL IN FULMAR VALLEY.

for a glacier tongue descending from the high ice-sheet west of the valley had so frequently changed its mind, advancing and retreating, that all the valley bottom and sides for about a mile were piled with its rubbish entangled with broken pieces of ice-foot. To make matters worse there was fossil ice beneath this moraine, preserved from melting by its stony cover. The mud and stones on the ice were in a loose condition, cracking in some places, moving viscously down steep slopes in others. It was rough walking everywhere, and, of course, most difficult for ponies. Rusty red puddles filled occasional hollows; torrents meandered about, tunnelling and emerging at the most unexpected points. A considerable river flowed along the bottom of the valley in a gorge of ice through the midst of this hideous chaos.

The survey kept me one side of the river; exigencies of the ground forced the caravan over to the other. They were advancing at last with fair speed over a snow-bed which, it was apparent from my position, would soon land them in difficulties, for it narrowed away to nothing and ended on an ice-slope, plunging direct into the river. The foolish Carl was leading the way, and on he went, with all the phlegmatic carelessness of a Balti coolie, till both he, the pony, and the sledge began sliding downhill. It was all that Garwood and Gregory could do to save them from perdition. After watching this incident from afar off I rounded over a last high mound of boulder-clay, and lo! a horrid view burst upon me. Hoping to see the pass close at hand, I beheld instead only the interminable valley slowly bending round and stretching away. If it had been desolate before, it now became of a yet more dreadful desolation, surpassing anything I ever saw or imagined. No view could be more simple. In the midst was a river flowing between banks of ice; on either hand long slopes of naked débris stretched up in unbroken sweep to a straight hill crest just edged with snow.

There were no buttresses, hardly any gullies, no precipices or emerging rocks, and no peaks above. The whole thing bent round in a slow curve. "Here indeed," I thought, "Nature ends." The worst feature in the view remains to be mentioned; it was a wall of ice that blocked the valley's head, the snout presumably of some great glacier we should have to surmount.

All the happy auguries of the previous day were disproved; the worst seemed to lie before us. I gazed upon the scene with fathomless disgust, but just then a gleam of sunshine broke through the wild black clouds, a golden gleam of the midnight sun, that glinted off the river and the ice-foot behind and gilded the brown moraine, making an island of glory in the midst of the bleak and dreary wilderness. A little snow-bunting settled beside me and cheerfully chirped his lark-like notes. It was the happiness of a moment. Darkness and gloom swiftly returned as I crossed the river by a snow bridge and rejoined my companions (16th July), just when they were quitting the moraine chaos and expecting an easier spell.

They were doomed to swift disappointment. Nowhere in the world can there be a worse patch of ground to travel over than what lay before us. It looked so smooth and easy—the gentle slopes, the neatly macadamised floor—but it was a deceit, a delusion, and a snare. The whole thing was a soft bog, a mere foundationless mixture of mud and stones, into which the beasts plunged deep, frequently sticking so fast that one had to be employed to haul the other out. We tried high up and we tried low down—it was the same thing. There was nothing solid anywhere for a hoof to rest on. It was a question of wading through the stiff mixture with hurried pace, for the longer you rested on any spot, the deeper you sank in. If matters could have been made worse, the rain that chose this moment for falling would have effected the result, but we were beyond caring even for a deluge. A hideous draught blew along the rut and drove the rain before it. There was practically no vegetation on the bog slopes, save here and there in patches, where wet lumps of brilliant-coloured moss, red, yellow, brown, and green, emphasised spots more than usually moist.

When the ponies were almost dead-beat, we reached the ice-foot below the glacier wall. It was surprisingly small for so great a glacier front, and small was the river flowing from the ice. There was something quite unusual about the glacier too. It had hardly any moraine, and it ended so abruptly, more like a wall of ice built across a valley than the end of an ice-river flowing into it. The ice-foot was in fact formed not by the glacier in front, but by a smaller one on our left. The great glacier contributed merely a few little cascades, whilst away to the right, round a corner, were indications of another stream coming in. We afterwards learned that this flows from a lake at the foot of yet another side glacier, which was wholly hidden from us by an intervening ridge. If we had gone up Turnback Glacier, and had steered our way correctly over the ice-sheet, it is down this glacier we should have come. The ponies were almost dropping with fatigue, so here camp had to be pitched on the first flat place. Dry places there were not. We set the tents up in a bog (370 feet).

GOOD GOING.

CHAPTER XII

THE IVORY GATE

I WAS too excited about the next stage of advance to await the process of tent-pitching, but went straight on up the hill to my left, round whose foot the glacier bent at its termination. It would be impossible to exaggerate the dejection of mind which at this time possessed us all. Obviously the glacier was a very large one. The probability was that we must mount to its head, find a pass over, and descend on the other side, carrying all our things and spending nights in the open in this hideous weather. No one said anything, and I set forth in grimmest humour. The way was over the ice-foot and a torrent too broad to be jumped and too deep in its ice channel to be waded. I turned the end of it where it emerged from its glacier and so reached the foot of my hill, the ascent of which lay up débris poised in a condition of more stable equilibrium than usual. Up I went, rounding gradually to the right to overlook the glacier. Suddenly, on turning a ridge, behold there was Agardh Bay before me and Wybe Jans Water beyond it, dotted over with speckles and streaks of ice, sparkling in sunlight, and yet farther off in bluest distance the hills of Edge Island, from Whale Point northward, with beds of creamy cloud lying on their crests, casting blue shadows. The way was quite easy before us, and the distance not great. The transformation of emotion within me was as swift as it was delightful. I yelled wildly down to camp and waved my cap before setting forward from Prospect Point (900 feet) to pursue the ascent of the long ridge whose back I had now gained.

With every upward step the view became more interesting and wonderful. It was not the snout, but only the side of the great glacier that opposed our advance. The peculiar appearance of the wall was thus explained. The snout descends to Agardh Bay and spreads around into a huge oval. We had only to climb the wall, an easy matter, cross the domed glacier and descend the other side, when a few miles of flat though doubtless boggy ground would alone intervene between us and the east coast. Our problem was solved. We had come straight towards our goal by the directest route. The way had been long and often unpleasant, but it was the right way, and does not admit of improvement.

The reaction from disgust and dejection to triumph and delight was an indescribable joy. Prospect Ridge led upward, and I could not choose but follow on. It became broad and of gentle slope. Shoulder succeeded shoulder, and each was a goal in turn. The view developed in all directions, and the wild clouds disclosed great domes of snow and, southward, peaks but half revealed. But it was ever to the blue fjord that the eye returned, the wide blue fjord, stretching so far away, with its ice floes, and the dark purple hills that framed it, and those remotest mountains beyond, with flushes of yellow light amongst them. Broader and flatter grew the ridge, its surface like the surface of a stony field, soft and level. I kept to the right edge of it, overlooking the great glacier—so white and wide, undulating down from broad snowy domes, one behind the other, on some of which the sun always shone. There was no proper névé. The highest and remotest domes were still of ice, naked save in hollows and on slopes where snow had drifted; for here the first frost turns the sopping snow to ice without need of long-continued pressure to change the structure of the mass. In the midst of the glacier at various levels were lakes of so beautiful a blue that the patches of blue sky above were not more full of colour and transparency.

The higher I rose the wilder and more furious blew the wind, carrying me along. Hundreds of fulmar petrels were in the air all about the ridge, apparently playing with the gale. It seemed as though the place belonged especially to them. Onward I went, for there was always another step in the ridge ahead. There was no snow on the ground, though lower places all about were thickly ice-draped. Wind keeps this exposed area free, and flowers grow upon it, especially poppies and the ubiquitous *Dryas Octopetala*. A single reindeer met me in the way and swiftly retreated. At last even the petrels were left behind. A long final slope led me to the top—a wide flat space whence cliffs dropped away on two sides to the great glacier's upper area, an apparently limitless extent of purest white—not flat, but ever rising to dome after dome, one beyond another, far as the eye could reach.

I halted, leaning against the gale, and with difficulty set up the plane-table, holding it with both hands and spreading wide the legs. My summit was higher than the Sassendal bluffs, and only surpassed by the snow domes. Unfortunately I had with me neither aneroid nor camera. Inland the storm was raging, and the dark clouds gathered and hurried along over or upon the hills. To the south were peaks more or less known to me. Some we saw from Fox Peak, others from Sticky Keep. They rise in a region of glacier. A rainbow was standing on the glacier foot, whose wide extension I could now well discern, with the serac crest that seemed to overhang the face toward Agardh Bay. All the region east of the watershed was much snowier than west. The land and low valleys about Agardh Bay were as snowy as Advent Vale a month before. One could not stay long at work in such a gale. It blew all warmth and feeling from the hands. On these occasions only habit enables a man to face working at all; mere will would not suffice. Each line drawn upon the map is won at a measurable cost of pain. When I could stand no more, I hurried under the cliff's shelter, and rubbed life back into fingers and arms. Then the view sank into the mind, never to be forgotten—one of the greatest and most memorable prospects I have ever beheld—oh! the glorious world, where man has no place and there is no sign of his handiwork, where Nature completes her own intentions unhindered and unhelped by him. Such pure snows no Alpine height presents, nor such pale-blue skies, nor that marvellous, remote, opalescent sea with its white flocks and its yet more distant shore. No Alpine outlook penetrates through such atmosphere, so mellow, so rich. The Arctic glory is a thing apart, wilder, rarer, and no less superb than the glory of any other region of this beautiful world.

Returning to camp was a fierce struggle, for now the gale was in my teeth; but the slope was with me, and I grew warm again in the contest. By the fluttering tents food was in an advanced stage of preparation. The herald of glad tidings is always well received. After some hours' sleep and a monumental breakfast we were ready to make our final rush for Agardh Bay. The main issue of the expedition was to be decided, and all looked forward to the coming hours with high hopes and fixed determinations. Sacks were loaded with provisions, cameras stored with films. Everything was in order save one detail. In the freezing wind on the lowest end of Prospect Ridge I had dropped unknowingly the screw that fixed my plane-table to its legs. The march had therefore to be begun with an ascent to find it, only a slight detour and not disadvantageous, seeing that the holding of a conclave on the question of route in a position overlooking the best part of the way thus became feasible. The missing screw was found, firmly trodden into the soft débris, a good omen, I thought, for the day's success. We were, in fact, in for a run of luck. The view was not so fine as on my previous visit, but it was clear enough for practical purposes, and over Agardh Bay it was most beautiful. While we gazed, a stream of sunlight poured upon the great white spread of glacier below, and the end of a rainbow again stood up over the ice. Edge Land was clearer, and sky and sea above and below it were striped in cloud and shadow of gold and blue. At 12.30 A.M. (July 17), exactly one month from the day we first sighted Spitsbergen, we made our final start, running down a steep débris slope, and enjoying a short but swift foot-glissade into the snow gully between the glacier and the hillside. The glacier was entered, without difficulty, by a trough between seracs, and the long passage over it commenced. As some one said, we were "on a *pukka* glacier at last, with rivulets on its surface, crevasses, moulins, and the whole Alpine bag of tricks." Yet there is really no glacier in the Alps at all like this Ivory Glacier, none so white, none so wide, none ending in so strange a manner; for when it quits its own

AGARDH BAY FROM THE IVORY GATE.

proper hollow and emerges upon the open meeting-place of valleys, it spreads abroad in all directions into a huge low dome, on an oval plan, about three miles in diameter, and some four hundred feet higher in the middle than around the edge. The whole of this domed extension of the Ivory Glacier has come down since 1870, for in that year Heuglin records that the Fulmar Valley stretched back, green and flat, from the head of Agardh Bay. Over this dome lay our route—a novel kind of pass. We afterwards named it the Ivory Gate, partly from a lovely section of veined ice that gave us pleasure and partly from some ivory gulls that frequented our camp. It was not a pass in the usual sense. The actual col is buried somewhere under the ice, and is probably about 400 feet above sea-level.

The surface of the glacier, as one made its acquaintance more minutely, was not like the surface of a glacier in the Alps. Winter snow, falling upon it, fills its hollows, and water percolates the mass, which ultimately freezes, producing areas of ice of a loose and crystalline character, ready to break up when trodden on into nut-sized granules or prisms. Beds of snow, sodden with water, covered large areas, and awaited the first frost to turn into ice. Crevassed patches were dry enough, but where there were no crevasses water lay about in a most uncomfortable fashion, and freezing slush had often to be waded through. But Garwood, our elected guide, led skilfully and reduced all such annoyances to a minimum.

We bore away to our right so as to mount to the top (646 feet) of the great terminal dome of the glacier, which an hour and a half of smart walking sufficed to reach; delay being caused, as the desired point was neared, by a multitude of crevasses, roofed with unstable archings of winter snow, which had to be carefully tested before they could be used for bridges. Our rope was of the frailest, odds and ends of cord pillaged from the tents at the last moment, and hitched together, for our proper climbing rope had long ago been made into sledge harness, and otherwise turned to mean uses. We advanced gingerly, therefore, in doubtful places, and only after much proving of the ground with axe-thrusts.

The view from the crown of the dome was indeed superb. The immense white foreground, curving away to every outlook, alone sufficed to give it singularity. In the whiteness were infinite grades of tone, for all the surface was rippled and broken by crevasses with blue edges. But it was away to the distance that the eye chiefly turned, especially to the east, to Agardh Bay and Wybe Jans Water, the gem of the prospect, and the goal of our toil. There, beyond the edge of the visible ice, lay the mud flat, neither land nor water, burnished bright over half its area, and becoming ever brighter farther away, till the swamps ceased and water reigned alone, water smooth with the utter calm of distance, bright with the reflection of the golden over-clouded sky, and speckled with dots and lines of ice and with great bergs stranded on its shores. This entrancing prospect was before us for the remainder of our way, at the end of an avenue of hills, terminating on the left in Mount Agardh, a prominence of some dignity of form, and on the right in a row of bluffs, buttressing a snowy area, above which rose a group of collared peaks, so called by us because of the thin vertical-sided beds of intrusive rock, which cut horizontally through them, near their summits, and form a protruding ring around the neck of each.

Forward we went, but now down hill, towards Agardh Bay, debating somewhat as to where the bay might reasonably be considered to end, and how far along the flat it was necessary for us to go in order to complete the crossing of Spitsbergen. In our heart of hearts there was perhaps some unexpressed misgiving as to whether the descent from

DESCENDING THE FACE OF IVORY GLACIER.

the glacier to the plain might not after all present difficulties hard to overcome. The gently-curving surface of the ice always hid the edge, so that we could not make any choice of point for quitting it. Quite suddenly we came to the edge. The glacier ended in a cliff of surprising steepness and 115 feet in height. A yard or two away a tongue of snow led down, its lower part curving out of sight. We followed it curiously and peered over. Another tongue of snow sloped up and almost met it. A short horizontal traverse across the steep ice-face was all that was needed to join the two. Garwood cut a few mighty steps into the ice, and over we went. A brief glissade landed us on the moraine (220 feet high), which was of a most peculiar character, like a ring of foot-hills below the ice-cliff. Its substance consisted of ice, deeply buried under a thick accumulation of stones, all small and rounded like the pebbles of a beach, which in fact they were, as the geologists will explain.

The exit from this peculiar region was made down a narrow valley, which twisted about, and permitted us to emerge on the almost flat valley-floor by a startlingly sudden transition. There was no ice-foot to this singular glacier, for the simple reason that there was hardly any water issuing from it. It has no great cavern vomiting forth a mighty torrent, but only here and there a driblet of a stream which trickles down the face and, after a short babbling race down a gently inclined shingle slope, oozes over the muddy plain. We too ran down the shingle, yelling and jodeling, in the very culmination of joy, for all the difficulties of our journey were overpassed, and all doubts solved.

There were streams to be waded at once—what of it?—little cared we for streams, though the stickiest of mud formed their beds. Through them we went, so as to get away from the glacier, and gain a good view of its monstrous end. The more we saw of it, the more we

THE TERMINAL MORAINE OF THE IVORY GLACIER.

admired our luck. We had actually chanced to hit the only point whence it was possible to descend without infinite difficulty and delay. At most points it could not be descended. At all other points it could only be descended by cutting a staircase down some hundred feet or more of perilously steep ice-wall. Descrying a little way off a small patch of dry ground, I made for it, through half-a-dozen streams, the others following delicately, carrying their boots and stockings, and lifting high their feet to give them, turn about, a momentary relief from the ice-cold water. On the flat we lunched on a varied assortment of Emergency tablets, biscuits, lime-juice nodules, bits of chocolate, ends of Darbishire's smoked Westphalian ham, and a tin of jam. We named the lunching place Darbishire's Ham, for the patch of dry land was ham-shaped.

Then we wandered various ways, geologists geologising, photographers photographing, surveyors surveying. Far out on the plain we went to a last gravelly island, fragment of an old raised beach, beyond which was nothing more but hybrid stuff, with a watery surface and a muddy substance, not strong enough to support, nor liquid enough to float anything—a nameless quagmire which gradually merged into the sea, whose stranded icebergs looked down upon it like little hills. The view was less fine than from the glacier, for the broad sea was now foreshortened into a line, and Edge Land had either sunk utterly out of sight or become wholly enveloped in clouds. There was hardly a living thing in sight; one glaucous gull came over, and an ivory gull, a few fulmar petrels and a snow bunting. There were the footprints of innumerable birds in the mud, and of reindeer on harder ground. That was all. Flowers grew sparsely where they had the chance to grow at all. Snow lay low on the hills around, almost down to sea-level, on north-facing slopes. Clouds played about the collar-peaks, and sometimes settled on Mount Agardh, but they rather added to than detracted from the picturesqueness of the scene, though we would gladly have been granted a clear prospect up an apparently deep and considerable valley that stretched inland to the west, and seemed to supply the largest affluent to the waters emptying into the bay. The largest rivers in Spitsbergen are not those that flow from the biggest glaciers, but those which carry off melted snow from bog and débris slopes, so that possibly this valley leads some way into the country. Its direction suggested that by it a traverse might be made to the head of Low Sound.

We spent in all three hours and a half on the east shore of Spitsbergen. It was long enough. By seven o'clock we were back at the moraine, and very soon began recrossing the glacier. The fine view was now behind us, and for the exhilaration of doubt and expectation was substituted the solid satisfaction of success, a more enduring but a less exciting emotion, and one by no means inconsistent with such substantial comforts as a hot meal and a warm camp. Towards these we stolidly trudged, finding crevasses a tedious bore, and snow-slush a nameless discomfort. On the highest point of the glacier-traverse (762 feet), a halt was made for light refreshments and a last view of Agardh Bay, but much of the glory had gone from it, for Edge Land was lost beneath bright low domes of cloud. The sky alone remained infinitely beautiful with its shining stripes of yellow, blue, and grey, redoubled in the level sea. The collar-hills were clear, and less picturesque for the clearness, which manifested their rather mean dimensions. Only the curdled surface of the enormous glacier retained its full dignity, and made its magnitude even the more impressively felt since we had learnt its width by the memorable process of walking across it. This was our last halt. The remainder of the way was accomplished without incident, and by ten in the morning we were at camp, awakening the drowsy Carl, and clamouring for food. The passage of the Ivory Gate of Spitsbergen was accomplished, and the first stage of the return. Downpourings of rain, that presently set in, were the final instance of the day's good luck, for they might have come sooner.

CHAPTER XIII

RETURN TO WATERFALL CAMP

HOURS of awful weather followed. Wind howled and rain was cast in sheets against our tents, which faithfully kept it out, but made plenty of noise in so doing. The doors were tightly closed. Little recked we of what went on without till the careful Gregory, looking forth, discovered our row of boots acting as so many rain-gauges. Boldly rushing out into the slush with naked feet, he emptied them and laid them on their sides, as boots in the open should always be laid, a rule the thoughtless Carl never could learn. It was a dreadful time for the ponies. To begin with, there was almost nothing for them to eat and no solid ground to stand upon. Their tracks afterwards enabled us to infer some of their adventures. Apparently, after taking a good look round and nibbling a bit here and there, becoming frequently bogged in the process, they agreed that the valley was intolerable, and set off to return. They kept along the river bank on the snow, picking their way not unskilfully, and making fair progress, till they were brought up sharply by lack of foresight. They were on an ice promontory, which fell precipitously to the river on one hand, a depth of twenty feet, and was cut off on the other side by impassable bog. A snow-bridge at the point of the promontory spanned a side stream that entered there. They considered this bridge maturely, advancing to it, putting forth a tentative hoof, retreating, advancing again, and so on, till the whole firm area was trodden down with their footprints.

Printed by Lemercier Paris

GLACIER FRONT IN EKMAN BAY

Ultimately, they gave up the adventure, and returned melancholy toward camp. At some other time they set forth straight up the hill, and were led by the harder ground into a stream-bed and up a gully, which they climbed to a considerable height. During these peregrinations, the hobbles were cut away by sharp rocks, for it is a peculiarity of such loose bogs that flat stones tend to stand upright out of them, like knife blades, so that a rope dragged about becomes easily cut. It happened that the only spare drag-ropes we had for them were their reins, which thus became lost. Our fruitless search for them revealed the story of the ponies' wanderings, and showed how energetically they must have moved about, for their tracks were everywhere. When the storm was at its worst, they came back to camp and took shelter close up against the tents, to our no small discomfort, for they pressed their heels or their noses against our dozing heads, and kept stumbling over the tent-ropes. Now there is nothing more disagreeable to a man in a tent than to have some one kick a tent-rope ; it jars the whole place, giving a peculiar sense of insecurity, and the process, if repeated, is as unpleasant as the accidental tapping on a man's tall hat of the points of the opened parasol of a lady sitting beside him. Spits's nose was continually searching in my neighbourhood for biscuits, and no persuasion would induce her to transfer her attentions elsewhere.

Notwithstanding all discomforts, it had been our intention to stay another day at this camp, and climb some neighbouring peak. There seemed little chance now of carrying out the plan. The outlook from the tent door was horrible. The foreground was a mere mass of sludge, so soft that the foot went ankle-deep into it the first step. It was all trodden about and puddled up into a filthy tangle, in which sledges, boots, tins, ropes, and all else were mashed together. Everything one touched was cased in mud. Mud covered our clothes, oozed over on to the tent floor, and gradually encroached everywhere, so that we became indescribably filthy. There was no opportunity to wash, for there was only liquid mud to wash in. Beyond a brief radius all was milk-white fog, through which the wind seemed to blow and the rain to fall without moving it. The fulmar petrel procession continued as before, and ghostly birds kept emerging and vanishing on their endless eastward way. Two brent geese were standing on the snow close by the tent, surrounded by their young ones. How the rain did pour down! There was not much encouragement to keep abroad. But the ponies needed consolation in the form of such small ration of oats as remained, eked out with a few biscuits. The brutes were dreadfully hungry, and snatched at the poor fragments with eagerness. They rubbed their heads against us, and seemed to beg to be taken away. At last, though the fog did not lift, the rain almost ceased. There was no more talk of peak-climbing; the word was, "All hands to the packing." Down came the tents, their canvas stiff and heavy with wet. They were rolled up anyhow, and layers of mud of necessity rolled in with them. It was a filthy business. Hands, and ultimately faces too, almost disappeared behind a brown mask. The extremity of our discomfort induced a strange indifference, wherein the discomfort itself almost disappeared. "There is," said Gregory, "a portage in Canada, called the Speak-no-bad-language Portage; I wonder whether it deserves the name better than this valley."

By slow degrees the sledges were loaded, laced, and roped with the foul apparatus, and the preliminary work was done. If the bogs were bad on the way up, what would now be their condition? This was the thought of each, but we cared little, for nothing could be worse than existing circumstances. Forward movement would be a relief, into whatever troubles it might lead. The ponies were unusually willing and co-operative from the start. I went off ahead to prove the ground, and Bergen followed, treading piously in my footsteps, and hardly ever hanging back. We took a new route, keeping down close by the river-side, where there were great snow beds and banks. Earlier in the season an arch of ice entirely covered the stream and extended a hundred yards or so on either side. The arch had broken in, and its abutments had in consequence crevassed longitudinally, as well as transversely at the entrance of the still over-arched side-streams. The transverse crevasses had to be crossed. They were not wide, but they were broader than a pony's stride, and their edges were rotten. The beasts, therefore, had to jump. As Garwood said, it was the Spitsbergen Grand National. Bergen's behaviour at the first jump was amusing. Observe that we never beat the ponies, we had nothing to beat them with, for the whip was voted a nuisance, and after being tied on to the sledge for a few marches was ignominiously left behind. The beasts lived with us, marched with us, and were like members of the party. They almost shared our tents, and they looked to us for their little luxuries of biscuits and sugar. Thus they had become as tame as dogs, and responded to vocal encouragement with remarkable confidence. I was on the far side of the crevasse, where I trod a firm landing-place. Bergen was on the other brink. There were bogs and starvation behind, hunger in the midst, and a chance of oats ahead. He deliberated, pawed a little on the snowy edge, and then jumped. The result was a triumphant success. Where Bergen went, Spits would follow. We found, too late, that their names ought to have been transposed. Other crevasses succeeded and were similarly passed; we thus turned nearly half the bog-slopes and some of the worst places. It was rather exciting, and we warmed to the work, but the outlook always remained the same, and rain kept falling at intervals.

It is remarkable how, when a man is at work in the open air, nature brings his mental condition into harmony with his surroundings. Whilst waiting somewhere for the caravan I had time to look about. Undoubtedly there was an element of grandeur in the scene. The fog dignified everything it did not hide. It magnified the ice-walls by the river into mighty cliffs, and broadened the snow-flats on either side into a weird wilderness, ending in large slopes of featureless bog, grandly curving and fading away into vagueness and final invisibility. By the water's edge the broken ice took strange forms, jutting up or overhanging after the manner of seracs.

When the snow ended there was no alternative but to traverse the bogs again. We followed our old tracks to the edge of the moraines and over them, with slight deviations where improvements were obvious. This moraine-traverse with sledges was really the most difficult piece of work we had in Spitsbergen. The place is not easy to describe. Let the reader acquainted with glaciers imagine a steep-sided main valley with a glacier snout entering it at right angles down a steep side valley. This glacier formerly blocked the main valley to a height of some four hundred feet. In retreating to its present dimensions it left behind its moraine-laden snout, through the midst of which the main valley's river has cut a gorge, still leaving piled up on either side the old moraine mounds with ice under them. We had to force a way over these moraines, which are perhaps best described as ice-hills, covered with deep mantles of mud and stones, and plastered on the steep hill-sides to a height of from four to five hundred feet above the main stream, whose immediate banks are walls of the old ice. The moraine covering was all wet and loose; sometimes there would be a slowly flowing mud mass many feet thick to be crossed, sometimes loosely crested ridges of stony stuff, all alike trending downwards, and so across the direction of our way.

The route actually followed cannot be described. It was up and down, to right and to left, over one obstacle and round another, but always forward, winning progress yard by yard. Sometimes the sledges had to be lifted by our united efforts, sometimes dropped over crumbling edges, sometimes convoyed down streams, sometimes hauled through pools. Rarely a small bed of snow or ice could be followed, and gave welcome relief. Half-way over, we were forced down to the river bank, where an ice-bed presented an irresistible attraction. It was of beautifully clear blue ice, frozen out of flood water and snow, in a small terrace-formation not unlike a miniature reproduction of the destroyed New Zealand terraces. Along this we gaily went, only to find farther progress impossible and return compulsory. The river must be crossed. There was no other way. Searching out a place where the ice-bank was low, and the torrent broad and of stony bed, in we plunged. The sledges floated at first, then got water-logged and rolled over. One entangled itself with a big boulder, and involved Gregory in a more thorough ducking than the rest. The far bank was steep and of loose rocks on ice. It took the ponies all they knew to climb on to it and haul up the loads. The remainder of the traverse was relatively easy, and who shall describe the joy with which we emerged on a mere stony river fan, and beheld beyond it the pile of goods which marked the site of Delusion Camp? The actual breadth of the moraine wilderness was about a mile. I did not observe how long it took to cross, but all agreed in estimating the time at about forty years, more or less.

I now relieved Garwood of the reins he had patiently and skilfully handled for so many days. My plane-table could be packed on the sledge. There was no surveying required. So Garwood and Gregory went ahead, when the loads had been rearranged, each armed with a geological hammer and marking their way with broken stones. I had a very soft billet, and won the credit of self-denial at the cheapest possible rate, for it was soon apparent that Bergen knew the way as well as I did, and could give me points in the matter of selecting the route in detail, so as to avoid banging the sledge against big rocks. Moreover, he was in a hurry to get down, and presently, to my astonishment, broke into a pathetic little trot over a flat place, as who should say, "Let us hurry up to quit this miserable place." Thus we made famous progress and left the fog behind, whilst the clouds soon lifted, and only remained on the crest of the hills at their normal roof-level of about 1500 feet. On we plodded through a known region. The valley had seemed a picture of desolation on our upward way; now it appeared fertile, kindly, and almost serene. The Sassendal was in sight, or rather the hills along its north bank, and with every hour the distance separating us from it visibly diminished. Yet the ponies showed no desire to halt. They kept pegging away like my old Balti coolies down the Baltoro Glacier, or Gregory's Zanzibaris when they had set foot on the Uganda road. Fatigue was simply not an element in the situation. All forgot to be tired.

Only when we emerged on the gentle slopes of the Sassendal did the question of camping arise. There were two conditions to be satisfied—a dry spot for the tents, and good grazing for the beasts. Bergen took the problem in hand and headed for a likely place, but neither he nor we were satisfied; other sites were in turn rejected, and the advance continued till we came to the valley that splits the Milne-Edwards bluff in two. We were on the point of halting. "Why not cross the river while we are wet?" was Garwood's wise question. So we limbered up again, plunged down the old raised beach's bluff, bumped over the stone-fan and through the rivers, and so reached the raised beach beyond. It was dry and flat, the conditions were satisfied, and Bergen agreed. The ponies were turned loose, camp (164 feet) was pitched, and supper followed. "Tired?" we asked one another. "Of course not; we are ready to make Bucking-horse Camp to-day if you like." But it was a curious thing, that though every one retired to his sleeping-bag with an elaborate apparatus of note-books and writing materials, not a note was written at that time. Slumber came unsought, and by general consent it was succeeded (July 19) by a Europe morning. I don't know whether some African germ revived in Gregory; however it came about, he was a little unwell, and thought he would try one of the admirable Pioneer Pills, which are the last kindness I confer on my best friends. "How are you this morning?" I asked him next day. "Oh, all right! those are famous pills of yours." "Did you take them?" "No! I didn't exactly take them, but I looked at them the last moment. It was enough. They are most rare pills."

We quitted Milne-Edwards Camp at the apparently reasonable hour of 10.40 A.M. (July 19). As a matter of fact this was very late, and marked the final loss of a whole twenty-four hours' day, for we had been gradually starting later and later, first about noon, then in the afternoon, then towards midnight, then early the following morning, and now we had come round to a normal hour. I encouraged this process because it saved provisions, involving only twenty breakfasts and twenty suppers in twenty-one days. Perhaps we overdid economy in provisions. We rationed ourselves on the most scientific principles, and the food we took abolished hunger at periodic intervals, but we began to doubt whether it was so sustaining as a generous diet of ordinary meats and other less scientifically combined preparations. At all events, for some not precisely discoverable cause, we were suffering from a peculiar slackness, and fatigue came quickly upon us. It may have been the fault of the weather, or the damp, or the climate, or the lack of sun. The air of Spitsbergen is not stimulating. It resembles that of a moist English spring, when the ground is clammy beneath a dripping sky.

So we lay late in bed, listening to the call of a purple sandpiper, or writing rhymes, and all the while it was really a vague reluctance to put on our wet stockings that was the most potent factor in postponing our start. Why one should have such an objection to wet stockings was a problem I could not solve. Ten minutes after the day's march began the driest hose must be as wet as the wettest, for stream-wading was a continual necessity. The fact however remains, and is here veraciously recorded. The day's march was a dull enough affair and long. The old tracks were practically retrodden, the ponies being left to their own devices much of the way and giving no trouble. Spits plodded steadily. Bergen's method was to make periodic halts and then trot on and catch up with his companion—the quaintest little trot imaginable.

It was wonderful how the rivers had gone down in our absence. Acres of channels were dry and the deep places were grown shallow,—a blissful change. Nevertheless there was water enough to pass, and the loads got so thoroughly wetted that, as Garwood said, the only dry thing left in his bundle was his Norwegian grammar. There were some bogs, of course, and equally of course Spits got fast in one, but this no longer disturbed either her equanimity or ours. As soon as she found herself well fixed she began complacently grazing on whatever green tufts were within reach. It was the finest feeding day the ponies had had; young grass was sprouting everywhere in delightful profusion, and at one spot was a crop almost long enough to be made into hay. In fact, it may be put on record here that about this date the valleys of central Spitsbergen become travellable, and that exploration into the interior cannot be carried out with much advantage at an earlier date. We reckoned that, assuming we had landed at Advent Bay about this time, and possessed of our present experience, we could have reached the Sassendal in three days and Agardh Bay in three more, provided, of course, that there was no surveying, geologising, or collecting to be done, and that the only problem was to cover ground.

On a direct line between the foot of Peak Milne-Edwards, which M. Rabot climbed in 1892, and Sassen Bay, where he landed, we picked up a peculiar piece of wood, very neatly turned, and shaped for some exact purpose we could not define. It was probably dropped by the Frenchmen. We carried it off as a trophy to be sent back to them. Shortly afterwards we saw a reindeer with a big pair of antlers, a great creature, double the size of the ordinary run of his fellows, and almost white in colour. With him was a dark-skinned buck of smaller dimensions. They were wary enough, and made off uphill before we came anywhere near them. These were the sole incidents of this most monotonous march, which was not enlivened by any discovery or novelty of view or effect. The cloud roof hung low upon the hills, and distances were blotted out. At Bucking-horse Camp we picked up the things that had been left there, and crossing the shrunken river below the dwindled waterfall, mounted to the knoll of Waterfall Camp, and set up our tents near the pile of fossils and empty tins which marked their former position. The march had only lasted five hours, and was not more than ten miles in length, yet we were all dog-tired, so that indeed I fell asleep while supper was in preparation, and waked to find my companions not much more alert.

THE SASSENDAL AND SASSEN BAY FROM STICKY KEEP.

CHAPTER XIV

WATERFALL CAMP TO ICE FJORD

IT was pleasant to awake amidst familiar surroundings (July 20), and find ourselves at a kind of home; pleasant too to be greeted by promising weather, with occasional sunshine and breaking clouds. The blue days of Waterfall Camp were come again to greet our return; indeed this was the brightest blue day of them all. With the advancing hours, tones became ever more lovely, more transparent and rich, more indescribably ethereal, so that, where deepest shadows lay, the substance behind them seemed most insubstantial. A quiet morning's work in camp resulted in the patching of the sledges, and the division and packing of all the baggage and geological specimens here hoarded during our eastward rush. Gregory was carpenter; his friends would have laughed to see him perspiring in the sunshine, with coat off, sleeves rolled up, and a tropical pith helmet on his head. How so unusual an object found its way to this spot I cannot say; it now appeared for the first time and suddenly, yet a pith helmet is not a thing a man can carry around concealed in his watch pocket. I asked for no explanation, preferring a world wherein mysteries survive to all possible worlds where everything is accounted for.

At two in the afternoon Gregory and I went off with two heavily-laden sledges for Sassen Bay, leaving Garwood behind to await Williamson's return with the ponies, which he was to convoy back to Advent Bay by Brent Pass. There was plenty of work for Garwood to do meanwhile, for important photographs were lacking, and a geologist apparently always can find sections to measure up, and other like occupations. As usual I went ahead alone, this time as far as Trevor-Battye's *cache*, having the wide valley, the sunshine, and the exhilarating wind all to myself. Of course I met a reindeer, a buck about three years old with horns in velvet. We advanced to inspect one another, and I sat down to give him an opportunity. He came up, grunting exactly like a pig, and seemed to prefer grazing close by me to wandering on, though whenever I moved he skipped awkwardly away. His nostrils moved up and down with great rapidity as he grazed.

At Delta River there was water to cross and a high raised beach to ascend, but all the rivers in this part had by now shrunk to moderate dimensions, and long boots kept the water out. This river emerges from the hills through a fine little gorge similar to that of its neighbour, the Esker. Its waters seemed to disappear entirely into bog on the plain, and only so to filter into the Sassen stream. This bog was rippled over with brilliant green waves of moss, with multifold convex fronts; between them were shining pools of water, gay in the sunshine, a beautiful decoration.

Henceforward Temple Mountain was ever before me, increasingly fine in the waxing afternoon, as the sun crept round it and peeped into its ravines, cutting out in bright lights and shadows its buttressed front, which sometimes resembles a columnar façade, sometimes a row of apsed chapels between the flying buttresses of a great Gothic nave. It is impossible to exaggerate or, for me, to define the quality of the beauty of the blue shadows that glorified the mountain this day. Deep, soft, transparent, ethereal, and yet, as it were, forming part of the mountain, manifesting the massive and monumental character of its structure,

GIPS BAY AND TEMPLE MOUNTAIN FROM SASSEN BAY.

but at the same time making it appear light and delicate, a faëry thing. Every step opened the eastward continuation of this mass of rock, where other cathedral-like promontories, with wide flat glaciers between, all tributaries of Post Glacier, carry on the architectural suggestion, resembling a group of splendid public buildings beside streets paved with whitest marble in the midst of a city of giants. Regarded as mountains they may in fact be small, but their regularity brings them into comparison with the works of man, and renders them strangely impressive.

Rubber boots are intolerable to walk in for long. I took mine off and stood barefoot. The ground was of springy bog, wet or dry, a delight and a surprise to tread on. I found walking thus a most agreeable experience. Incipient fatigue swiftly disappeared. I cared no longer whether I sank into bog or not. It was a new delight to feel the play of all the muscles of the foot, and how it spread abroad like a hand and responded to every inequality. There was no need to tug it from the clasp of bog or mud; it came out easily, and the softness under foot was no longer a nuisance, but a pleasure. The cold ground and cool air felt fresh and kindly. I had found a delightful new sensation. Three or four miles were thus covered before stony places returned and boots had to be resumed. I have seen men in the East, who habitually walk barefoot, go most gingerly over stony patches. Henceforward this will cause me no surprise.

Beyond the last river was yet another raised beach, from the point of which Sassen Bay was visible close at hand. The ponies were cast loose to graze. It seemed that the march was done in shorter time than was expected. Gregory told me we must either camp at Starvation Bluff or Windy Point—a sorry-sounding alternative! Starvation Bluff is the natural landing-place for Sassen Bay; it is a low raised point close to a pool of clear water by the cricket-field—the piece of flat dry bog where, I am told, the passengers of a tourist steamer once played a match. Nature has provided a table for the scorer, in the form of a weathered rock on a neat pedestal that stands out of the grass. We used to talk of the place by the name Starvation Bluff, because it was there Gregory and Pedersen spent hungry hours awaiting Trevor-Battye, what time the bay was blocked with ice. Afterwards Trevor-Battye landed here and left tents and supplies for use on our return. We found them near the water's edge, but no boat to enable us to reach Post Glacier, as we intended, or to communicate with Advent Bay.

There was now little ice about; only a few white and blue masses, large and small, fallen from Post Glacier, and either floating in the bay or stranded by its shores, objects most beautiful in themselves, and a charming addition to the glories of the landscape. The sky was clear, save for a few cirrus wisps; the sun shone brightly over the bay. A chilling wind blew, but I escaped it by jumping down the little limestone cliff forming the face of the bluff, where a tiny cave about four feet high was all ready for me, filled with sunshine, and no breath stirring. The rippling waters plashed and broke in wavelets a yard away, strewing a line of fresh, sweet-smelling seaweed along the shore. It was good to be down by the sea again, the sea that unites all the countries of the world, and brings mankind together. No human creature was in sight, but a few pieces of shaped driftwood abolished the sense of utter isolation. We were back by an element on which men live out their lives; we had left behind the region where life is only temporarily supportable. These sentences of course contradict other opinions expressed above. Such changes of feeling in varying circumstances form part of the charm of travel, and much of its profit. They open the mind to new sympathies,

POST GLACIER AND TEMPLE BAY FROM SASSEN BAY.

and force it to adopt varying points of view. Now it seemed good to behold the flocks of eiders swimming on the waters, and the petrels flying swiftly just above the surface. The guillemots were like old friends found again, and we even forgave a glaucous gull, feeding close inshore amongst a long line of kittiwakes, for the inland misdeeds of his fellows.

Sassen Bay, beheld from this point, is not only worth seeing, but worth coming so far to see. Here the Temple and its companions are at their grandest, with the blue fjord or the white glaciers at their base. Hence, too, one beholds all the wonder of Post Glacier, with its splendid shining width, its incredible levelness, its noble backward sweep, and its finely-curving tributaries, smooth and wide like itself. It ends in the bay in a wall of ice, a wall white and blue, splintering into a broken edge. "Oh to be on it!" was an immediate impulse, as usual needing the correction, which I suppose life is not long enough to make habitual, "Dort wo Du nicht bist dort ist das Glück." Westward was the long reach of Ice Fjord, closed by the pointed and contrasting row of peaks bounding the great glaciers of Cape Boheman, behind which stood up a tall and remoter pyramidal white mountain, a little Weisshorn, which was playing the pretty game of being in shadow when its neighbours were bright, and then shining forth brilliantly when a cloud came between them and the sun.

Altogether, Starvation Bluff seemed a charming spot for a camp, but a plum-pudding determined us to go on to Windy Point, a few miles farther along the coast in the direction of Advent Bay. When Gregory came back after his long walk round to our base, he was obliged by the gale to land at Windy Point, and there accordingly he left the stores intended for use on our return. Thus we had this day to pick up Battye's *cache* near Delta River, the tents and things left by him at Starvation Point, and finally, if we wished to make a complete collection, Gregory's deposit at Windy Point. In the last was the plum-pudding, which decided the issue.

The sledges were thus indeed heavily burdened, and to make matters worse they had to be dragged over slopes furrowed with innumerable gullies. I only beheld their troubles from afar, for work kept me close to the beach. Gregory toiled exceedingly to help his beast, constantly having to lift the sledge-end on to banks, whilst the ponies became tired, and unwilling. When, at the last, a ridge had to be mounted to a height of perhaps 300 feet, the labour for every one was almost too much. My walk along the beach would have been delightful but for the cutting wind. There was always the beauty of the fjord, to which the charm and interest of the coast were added. A long overhanging snow-front still remained in many places, undermined by the tide, and breaking away in great lumps, soon to disappear. The beach was covered with traces of man, an old milk-tin ("made in Germany"), a quantity of tent-pegs white with age, driftwood of shaped fragments, besides large sawn pieces of timber never intended to find their way to these shores. Windy Point is the jutting raised delta of a glacier stream that comes from Flower Valley. We followed along it to close under the hills, where was a wooden signal, set up recently for survey purposes by Baron de Geer; near by were the traces of his camp. Here accordingly we halted and set up a noble camp of three tents, our two large ones from Advent Bay and the little up-country tent for Williamson. The work of tent-pitching went much against the grain, but was ultimately accomplished about half-an-hour after midnight. We were too weary to be able to take much profit from the wealth of supplies, but went soon to sleep, promising ourselves a copious reward next day.

SASSEN BAY.

At eight o'clock (July 21) Gregory roused me, and we wondered at the long sleep of about eighteen hours we had enjoyed without a moment's interruption. Just then the sun broke through clouds in the south-east, and we discovered it was eight in the morning! Our sleep had been for six hours only. In due season Williamson went off with the beasts and empty sledges to pick up Garwood's camp and fossils at Waterfall Camp, and follow him to Brent Pass, where they were to meet.

Garwood spent the 20th collecting fossil reptiles from the Triassic rocks on the Trident. He returned to Waterfall Camp at one A.M. on the 21st. His journal describes him as lonely, tired, and seedy, with damp in his inside. After a sleep he packed up his tent, left it for Carl to bring on, and walked over Brent Pass to the head of Advent Vale.

"I was accompanied," he writes, "by a snow bunting, who hopped beside me for half a mile, without the slightest trace of fear, singing his sweet little song while I whistled an obligato. As I waded the countless channels, which rippled over the moraine-strewn flat at the foot of Booming Glacier, I realised that in order to penetrate as far as possible to the south, it would be better not to ascend the Baldhead, but to follow a higher ridge, conducting farther east into the heart of the glacier system, whence Booming Glacier takes its rise. Stumbling over the loose moraine and wading ankle-deep in liquid clay, I reached the foot of a smaller glacier coming down from the south-west to which we gave the name Baldhead Glacier. This was once tributary to the larger glacier; it formed the old lateral moraine by which I had approached. Baldhead Glacier had now shrunk back into its own valley, leaving a narrow passage between its own retreating snout and the overhanging side of its advancing neighbour. A glance at the piles of ice recently fallen across this passage, decided me to keep up the lateral moraine, and cross the smaller glacier some distance above its snout. The moraine was of the loosest and most villainous composition, being, in fact, nothing more nor less than boulder-clay in process of formation. Reaching the left side of the glacier, I had some difficulty in finding a place where I could safely cross it, as the surface drainage, which flowed to this side, had excavated a channel for itself along the side of the ice. The channel where it was open was too wide to be jumped, and there was nothing to land on, on the other side, but a steep ice-wall; where the channel was roofed in, the snow was too treacherous to be trusted. By following it up some distance, however, I was enabled to cross. A similar difficulty had to be overcome in descending the glacier on the other side.

"On starting up the ridge I discovered that I had by no means exhausted the dilapidated peaks on the island. The arête was simply a pile of débris, but the size of the fragments was astonishing. I could scarcely lift most of them. They consisted of large slabs of yellow sandstone, many two or three feet square and over an inch thick, piled in a loose heap as though just tipped out of a cart. During the whole ascent I never found any solid rock. If I endeavoured to trace a fossiliferous fragment to its parent rock, by removing the loose slabs, I had eventually to give up the quest in despair, fearing lest, if I continued to remove everything that was loose, I should have no mountain left whereof to chronicle the ascent. Indeed, the answer of the Irishman, in reply to a tourist who had asked the height of a hill, that it was 3000 feet to go up, and 1000 feet to come down, would have applied well to Booming Peak.

"Portions of the arête were covered with snow, and afforded a welcome relief, though in places they were so steep and hard as to require step-cutting. The ridge was dull and tedious to climb, but afforded me the view I sought.

GAP BETWEEN BALDHEAD AND BOOMING GLACIERS.

It commanded the whole of the basin and feeders of Booming and Baldhead Glaciers. At my feet stretched the curious and interesting surface of the main ice-stream, with its swollen and serrated margin rising high above the centre, and resembling nothing so much as the row of finger biscuits surrounding a bowl of trifle. It was most interesting to observe how the glacier was advancing and forcing its way up over its old lateral moraine. The more rapid motion of the upper layers of ice caused the sides of the glacier to overhang in perpendicular cliffs, from the top of which ice-avalanches were perpetually crashing on to the floor of the valley below. This raised and broken edge could be traced for some distance up both sides of the glacier, presenting a marked contrast to the sloping, convex snout of the retreating Baldhead tributary. This is not the place to discuss the present climatic changes of Spitsbergen, but it may be noted that from this glacier we have evidence that the general advance shown by so many of the Spitsbergen ice-streams does not necessarily prove an intensification of glacial conditions in these latitudes.

"After descending to the foot of the ridge, I debated whether to recross the Baldhead Glacier or risk the short and easier passage between the two glaciers. Tired and hungry, I decided in favour of the quicker route, and, skirting close round at the foot of the smaller snout, scrambled over the fallen blocks of ice until I was forced to climb on to the old lateral moraine to avoid walking under overhanging séracs, threatening to fall at any minute. On the moraine I found a deserted eider's nest. The brood being hatched, I pocketed the down, thinking it might protect some of my bones when the inevitable fall came, without which I had little hope of clearing that execrable moraine. However, after wading the interminable channels on the flat below, I reached Carl and the ponies at three A.M. without mishap.

"Carl was asleep, and the ponies had strayed out of sight, being as usual, when left with that master mind, insecurely tethered. Indeed, having long since used up both pairs of reins and all our light Alpine rope for this purpose, he had probably nothing left with which to fasten them up."

Next afternoon they set forth down Advent Vale, going together as far as Ooze Camp, where Garwood bore to the left round the mounded hill so prominently seen from Cairn Camp. He passed over a low col, and went down an easy open valley on the other side to the site of Cairn Camp. He waited an hour for Carl, but saw no sign of him, so, to quote from his journal :—

"I started up the valley to meet him, and found him in difficulties with Bergen, who was tired, and had rubbed his shoulder. We struggled along to a dry patch below Cairn Camp and pitched the tent. I then discovered that my ice-axe and hammer had been dropped off one sledge, so we had to pitch the tent with my camera legs; also I had dropped my knife by the way whilst digging up plants, so it was indeed a day of misfortunes. Turned in about four A.M. (23rd)—raining.

"Started at two P.M. The travelling was bad from the start. We both fell into a hole in crossing Fox Valley River and were soused. There were alternately areas of large stones and swamps. Only on the terraces after passing Bolter Valley could we get firm ground for more than a few yards. In fact, in some respects, things were in a worse condition than when we came up, especially for the sledges, which were bumped to pieces over the boulder-strewn patches. Our worst difficulty, however, was the second gully, which, instead of being choked with snow, was now a deep trench with almost perpendicular walls, difficult for a man, but impossible for ponies. We had, therefore, to keep up the gully over fiendish piles of rock-débris, and eventually, in order to cross, up an icy slope of about 40°. Harnessing both ponies to one sledge at a time, and ourselves hauling with all our might, we managed to get both sledges up, but they were a wreck of their former selves. We had to keep high up to cross the next gully, and then down the steep upper slopes, where the native sledge utterly collapsed. First one runner gave way, then the other. The sledge was running on its bottom. Then that gave way. So I abandoned it to its fate, and harnessed both ponies to the remaining sledge, which was in the dickiest condition. We eventually struggled into camp at midnight, wet, tired, and miserable. It had rained and sleeted nearly all day, and I was several times in despair of getting either sledge through. Carl fetched in the broken sledge next day."

CHAPTER XV

MOUNT LUSITANIA

JULY 21 was a raw and mainly sunless day at Windy Point, but it was not unpleasant, for rain held off and clouds kept above the hills, so that the forms of the landscape were visible, though the previous day's wondrous charm of colour was lacking. Gregory did some fossil-gathering on the slopes behind, whilst I cooked; otherwise neither of us left the sacred precincts of the camp. Our only visitors were birds. Most welcome were the kittiwakes, who fed in a row at high tide a few yards from the shore, one and another now and again rising on easy wing and hovering over the water before lightly settling back upon its surface. A skua or two came to see that we were all right, and there were eiders and glaucous gulls and guillemots. Clouds grew lower and denser with the passing hours, and all the glory of the view was taken away. The Temple looked like a mere hump of rock, the glaciers like so much ordinary ice, and the bay like a sheet of water. Fancy sleeps in chilly seasons, and needs the sun to awaken her.

This time it was not the sun which came as awakener, but the spirit-lamp. It took some wild-freak into its interior and suddenly erupted, pouring forth a volume of flame, and ejecting a quantity of lighted spirit on to my right hand and in the immediate neighbourhood of two pounds of blasting-powder. I plunged the burning member into a sleeping-bag, whilst Gregory extinguished the volcano.

LOOKING SOUTH-WEST FROM MOUNT LUSITANIA.

Butter and an old duster were the only medicaments and bandage available, but they sufficed. It was a sorry outlook and a painful experience, but things turned out all right. I was not crippled nor prevented from holding pen and pencil for a day, thanks to the horny hide produced by a month's camp life. Spirit could burn on it as on the outside of a boot. It raised indeed a score of blisters, but they were judiciously scattered.

With early morning (July 22) the sun returned, and the wind dropped; hills were again clear, and waters smooth. As we sat over breakfast at the tent door a flock of little auks came to amuse us. They appeared to be young birds at school. There must have been a couple of score of them swimming about in a tightly-packed group, and turning hither and thither in unanimous evolutions. At some note of command given, down they all simultaneously dived, remaining under for about half a minute, and returning to the surface widely scattered, but swiftly concentrating again. Gregory had darning to do. I watched him. He was my superior in this matter also, for as a child he had been taught the art, much to his then disgust. In reply to his protests his mother gave him a book of Moffat's to read, in which that African traveller relates the trouble he was in, when thrown on his own resources, through not having learnt to sew. It was a curious coincidence that sent Gregory also to Africa to apply his youthfully acquired skill. This day he would have none of my exceptional apparatus, and sneered at my six-inch needles and ice-axe thimble, selecting only from my store a quite ordinary darning-needle and commonplace thimble—such is the narrowing effect of prejudice early acquired!

At eleven A.M. we set forth for a day's exploring inland, congratulating ourselves on the kindly weather. Our way was up the valley behind—the Flower Valley we named it, after the

Director of the British Museum (Natural History), and because of its relative botanical wealth. A naval officer once told me a story of his first Arctic experience. He was sent from his ship with a few men to land at some spot on the neighbouring coast and find a cairn built there by a previous expedition. One of the seamen with him was an old Arctic hand. When they had landed, the seaman said to him, "Where's your food, sir?" He replied, "I haven't brought any; we shall be back on board in three hours." "Excuse me, sir," was the reply, "but that ain't Arctic." "Where's your rifle, sir?" was the next question. "I haven't brought one." "That ain't Arctic either, sir." They climbed the hill and found the cairn. While it was being searched the seaman carved the officer's name on one side of a piece of wood, and his own initials on the other side. "What's that you're sticking into the cairn, Jack?" "Just a piece of wood with your name on it, sir." "But I don't want my name carved all over the place." "Excuse me, sir, but that's Arctic." Returning to the shore they found the ice packing, and were only just able by the narrowest shave to get back to the ship before the way was blocked.

We went forth this day without either rifle or provisions, except a few biscuits. As Gregory afterwards said, "It ain't Arctic." We needed both—the rifle when we found reindeer about half a mile from camp, and the provisions a few hours later. We followed the left bank of Flower Valley by an up-and-down way over various steep-flanked gullies. The scenery was tame, and resembled that of some of the less interesting valleys of Tirol. The slopes were bare, with snow-beds lying on them, and there were large snow accumulations by and overhanging the river. Masses fell from these into the stream with the boom of an avalanche, and cast clouds of snow-dust into the air.

The sun was hot ahead, and we felt his power, and perspired as much going up as we were to shiver coming down. The view over Sassen Bay developed with the ascent, but it lacked picturesqueness, because the sun shone straight on to the hills, so that there were no shadows to define and diversify their forms. Even the Temple looked flat and mean under such illumination. On the east was a line of hills, with an uninterestingly undulating crest; over against them was another similar line; the quiet valley lay between. About four miles up was the glacier's snout (928 feet), amongst a moraine chaos. There was a peculiar object about fifty yards below the end of the glacier. It resembled a small volcano, about forty yards wide, with a crater in the midst, the diameter of which was about twenty yards, and its depth, say, twenty feet. But the thing was made of ice, for the most part in long blue prisms, extraordinarily beautiful, whilst the crater contained a pool of water. It was like the place found by the Wandering Jew, where "the cracking earth shot forth bundles of frozen spangles." The pranks played by the ice-foot below a glacier's end in these parts are a constant surprise to any one only acquainted with glacier phenomena in the temperate regions of the globe. Any pile of snow that endures over one summer turns into ice. This mound was doubtless originally a pile of snow, perhaps drifted up by the eddying of winds in the moraine hollow in the midst of which it lay. The formation of a pool of water in it would not be surprising, and the pool once formed would tend to deepen and enlarge. It would be interesting to know whether this object always exists, or whether it is only a temporary phenomenon. As it lies so near the frequently visited Sassen Bay, perhaps future visitors may bring back news or, better still, photographs of it in years to come.

Crossing the river by a frail ice-bridge, we mounted the moraines and entered on the glacier. It was in no sense a peculiar glacier, but may be taken as absolutely typical of the smaller Spitsbergen examples. From the moment of treading on it you know that you are not in the Alps. The surface of the ice is softer, and crunches more easily than in Europe; there is more water on it, and more snow, lying about in wet beds, and now rapidly disappearing. This snow was in better condition than any we had thus far met, and had evidently been frozen during recent days. There was not much more snow high up than low down. There was no proper névé. At the col and right up to the tops of the peaks around, all was ice below the thin snow-sheet. The main water-channel started from the very col itself and flowed over the ice, even crossing snow-filled crevasses, right down to the moraines. These surface rivers, flowing in channels down the snow-covering of a glacier from its uppermost region, are highly characteristic of Arctic glaciers. We saw them even on the largest and remotest snow-fields. They often become bogged in snow-beds, or flow for a while below the surface, but they emerge again and continue their irregular furrow down to each glacier's foot. They form a climber's great impediment, for on large glaciers they are deep and swift, and the snow-bogs and slush-lakes they make are enormous, and may often be impassable.

As the day advanced and the sun went round, the widening view over the fjord grew in beauty. It strikingly resembled many a spring-time prospect over Como or Maggiore. Hills and water were alike drowned in a beautiful blue tint, which masked the barrenness of the shores whereon the sun shone warmly. No lake-view of Italy could have been more luxuriant or more lovely than that of this desert sea-reach in an Arctic land. But clouds were gathering over it when we gained the col, and the moments of its visibility were evidently to be few; so we did not halt, but hurried

FLOWER PASS.

on, up the rounded ridge of gritty débris on our left to the summit of the neighbouring Mount Lusitania, the self-same mountain that Garwood and I were making for on July 6, when the crevasses turned us back. The first thing I saw from the top was the place of our enforced return. The snow in which we so deeply waded was gone from it, and the ice-surface revealed with its multitude of schrunds. I gazed at the place with retrospective horror. A more crevassed piece of glacier I never saw ; and we had no rope. In future I shall say of unroped glacier-walking in July, "That ain't Arctic."

The view was important and interesting. There were two hours' topographical work in it at the least ; but no one could have stayed two hours to do it, for a cold and furious wind made every moment a torture, sucking all the warmth out of us, and swiftly taking every particle of feeling out of the fingers, so that a little longer exposure would have resulted in frost-bite. The time of our halt on the top was in every respect horrible. I levelled the plane-table on the loose flags, and then knocked it crooked when I moved, for the flags on which we trod tilted one another, and the work was to do again. In attempting to photograph I fogged three plates out of eight through the clumsiness of cold fingers and stupidity induced by the raging elements. Moreover, there was need for haste anyhow, for clouds were rapidly covering the hills, and one point disappeared after another. It was infinitely annoying, for we were in the very centre of the surveyed region, and could have corrected and added many details. The heads of the complicated network of valleys, whose mouths alone clouds had permitted us to see on the way up Advent Vale, were all now revealed. We were behind the outer ring of peaks that look down on that valley, and the secrets of their structure were revealed. There was the Sassendal in sight too, right up to Rabōt Glacier, and all the maze of puzzling mountains round from it to the Advent Vale Bluffs, with the Baldhead right in the foreground—work in them alone for an hour, but before we had well realised what they were, cloud covered them up. Then there was the splendid fjord, away to Cape Boheman, with countless hills beyond it—low ones near the coast, and isolated big fellows behind—the Colosseum, the Capitol, and unnamed giants far inland.

We stamped and slapped our arms about, and scrawled crooked notes, and took hurried sights and photographs—but it was all of little use; one could not quietly reason things out with frozen body and fingers recently roasted, and now submitted to a cold-storage process. For half-an-hour we suffered and laboured, then made a wild bolt for the pass, and left the storm to rage amongst the senseless rocks. Down we went by a shorter route, plunging again over concealed crevasses, into one of which I slipped a foot. It was fortunately small, and we were in no temper to bother about trifles. The wind did not reach far down. On the leveller glacier it was only a slow, cold air-current, and by the moraines it was almost still. There at last we halted under the lee of a big rock, ate our biscuits, and wished for something more solid. Continuing the descent we found better going on the right bank of the valley for some distance, then crossed the river by lucky stepping-stones, and so ultimately returned to camp with dry feet, for the first time in Spitsbergen, after a most fatiguing expedition. Never did the shelter of tents seem more delightful, or a hot meal more soul-satisfying. When the canvas doors were tightly closed and the pipe of solace was lit—in which Gregory knew not to share—little cared I that the weather was again broken and the future utterly obscure.

Next morning at breakfast we had a momentary vision of release; we thought there was a boat off Hyperite Hat, coming with a fair breeze to carry us away. Then we thought it looked through the glass like a steamer. But it remained always in the same position, and proved to be one of Baron de Geer's survey signals, raised and magnified by a tantalising mirage. Fog and rain presently settled down, and the hours had to be spent on camp work, of which there was always plenty to be done. The length of time one has to spend over cooking is alone a considerable factor in servantless camp-life, and the fewer members there are in a party, the longer is the proportion of time thus occupied, since four are cooked for as quickly as two, or more quickly, with more hands to do the odd jobs of fetching water, washing up, finding tins and things from the stores, and so on.

The day wore slowly out. A real sail appeared in sight, apparently coming from Klaas Billen Bay. This time we felt sure it was Trevor-Battye, but the boat, in which were two men, did not come in our direction, and was lost from view somewhere in Temple Bay. They were having a brisk sail of it, for there was a hatful of wind and white-caps all over the fjord. Clouds hurried by, low down, and rain-besoms swept sea and land below them. Towards evening rain turned to snow, which even fell about the tent for an hour, but did not lie. Through cloud-breaks we saw now this hill, now another, powdered white from base to crest, the Temple being selected for special adornment.

Tightly we closed the tent doors, whilst I for my part spent the evening hours seated in a warm fur bag, and indifferent to what went on without. The wind might howl itself hoarse, and the rain ding and splutter on the tent, I cared not at all, but wrote and drew with absolute indifference. Such treatment takes the heart out of any weather, however bad. Improved conditions consequently set in, and the following day (July 24) was bright again. Spitsbergen weather in these summer months is very English—the same soft, damp air, the same fickleness and unreliability, the same occasional perfection. In future I shall always think of England as belonging in a climatic sense to the polar regions. The Arctic Circle ought to be drawn through the Straits of Dover. The contrast between London and Paris weather is the contrast between the Arctic and temperate regions. Our fogs and winds and changeful damps belong to the pole. Our green lawns are but more refined Spitsbergen bogs. One has to come to these islands of the north to understand not merely the geological history, but the present atmospheric conditions of the British Isles.

Gregory and I were on the point of sallying forth to seek fossils and adventures, when a reindeer came down from Flower Valley, and put us in the way of a temptation to which we promptly yielded. It was not any sporting enthusiasm that awoke in us, but sheer greediness, the desire for a change of diet, for fresh meat to get our teeth into. The lot fell upon Gregory, and off he went for a stalk, the details of which were hidden from me by a fold of ground. It suffices to say that the beast was persuaded into the pot. May I never have a more toilsome job than the carrying him between us to camp, for he was a lusty buck. Whilst Gregory skinned him and cut him up, I went for a belated walk along the shore, keeping to the loose and deeply furrowed hill-sides, for the tide was high, and washed the little precipice at their foot.

It was a toilsome promenade, but not dull, for there was always the beautiful loch on my right, with the fair hills beyond, a harmony in silver and blue. About three miles along, the hills ended, and De Geer Valley debouched on a wide, far-extending delta, low and damp, with many streams and pools. Here was another of De Geer's camps, and several of his signals about, among them our quondam deceiver. The valley rose by a sudden step over a wall

LOOKING UP DE GEER VALLEY.

of hyperite. On to this I mounted, and the whole trough was open before me, long and straight, leading to a low green pass, beyond which was the valley, which opens into Advent Vale at the old moraines. On my left was the dull range forming also the left side of Flower Valley. On my right were a new set of hills. The valley was wide, fertile (for this region), and flat. The ponies might have returned by this route to Advent Vale, and been saved a long detour. One cannot, however, foresee everything in an unsurveyed country. I advanced some distance up the valley without gaining more information.

Returning, I was attracted by a faint cloud rising over the river, where it flows through the hyperite. This demanded and rewarded investigation, for it was caused by a striking waterfall. The hyperite gorge has vertical sides, whose effect of height is increased by the roughly columnar structure of the rock. Half-way along the gorge the river plunges over a wall about fifty feet high, itself transverse to two longitudinal walls of double the height, enclosing both the cascade and the pool into which it plunges. In the afternoon, when I was there, the whole was wrapped in shadow, but the gloom magnified the aspect of the scene. One or two rock columns stood out, almost detached from the masses on either side, great monoliths finely erect. The gorge, bending round behind, closed in the view. Though the fall and its surroundings are not on a large scale, the proportions are excellent, designed on noble lines, and more impressive in total effect than many a greater assemblage of rocks and falling water.

The light was bad for photography, but I could not wait the needful five hours for an improvement. My steps turned homeward, or rather dinnerward. The tide was out now, so the narrow beach could be followed, an agreeable change from the everlasting up and down of gully-crossing on débris slopes. Little waves came tumbling in rapid succession on the fallen blocks of hyperite and the fine débris of black shales that lay between the low shale cliffs and the water. Overhanging ice-masses still clung to these cliffs, all undercut through their whole depth of twenty feet or so, and merely clinging to the cliffs by sheer adhesion. Water flowing from the gullies was discharged through ice-tunnels that invited one to enter their dark twisting recesses. At one place a fulmar petrel, apparently injured, flopped along the ground at my feet so that I could touch it. It made its way into the sea and swam leisurely, about two or three yards from shore, and quite unconcerned at my presence. There were lumps of clear drifted ice on the beach. Near camp the hyperite blocks became so numerous that it was necessary to step from one to another, and walking grew wearisome, as it always becomes when the eyes have to be fixed on the ground. Arrival in camp was therefore a relief. I had only been absent five hours, but lonely tramps always seem long. This walk, however, had not been unprofitable, for it enabled me to fill a blank of twelve square miles on the map. The vacant space between Advent Vale and the sea was rapidly diminishing. One day with a boat would cause it to disappear. Before turning in we again scanned the horizon with a glass, but there was no sail. "To-morrow," we said, "they will assuredly come and fetch us."

HYPERITE WATERFALL.

CHAPTER XVI

BY SASSEN BAY

ON awaking (July 25) we looked forth. There was no boat, but there was a fox making free with our meat at the very door of the tent. He was a mighty cool animal, and regarded us with contempt. Unfortunately for him his skin was needed for the British Museum. He dragged a joint a few yards off, and remained by it while Gregory fetched and loaded the rifle which put a period to his existence. A shot-gun would have been more suitable, but there was none in camp, and the only rifle was an old Snider, that had been banged about in African camps. The fur, of course, suffered.

Gregory skinned the beast, whilst I scanned the fjord for a sail and smoked many pipes. Deceptive lumps of ice hove in sight and disappeared, but never a sail did we see. The sun moved round in the clearest sky ; a cool breeze played about the tents. The sea was soundlessly still, only the neighbouring burn sent a rippling voice through the bright air. It was a day amongst a thousand, delicious to every sense, and perfect in beauty. Some hours went uncounted by in blissful contemplation, and I, for one, was looking forward to a whole happy, vacuous day. It is easy enough to promise one's self the enjoyment of leisure hours in perfect idleness. The guests of Nature, as all men are, cannot reckon without their host, and she abhors a vacuum, nor long permits even a traveller to be happy unoccupied. So when the afternoon came and brought no boat, by mutual consent we quitted camp for an upward ramble.

The tents were planted between the fjord and a corrie running back into the wide hill-front, the end of the ridge separating Flower and De Geer Valleys. It was clear from the survey that this ridge must be wide, and should have an undulating top, whence fine views might be obtained. Accordingly we made for the corrie's mouth, and were surprised to encounter not merely a moraine, but one still covering accumulations of old ice. The glacier that left the ice has long ago receded and, as we supposed, utterly vanished. We found it again, however, as will presently appear. The crest of the corrie, all round, as seen from below against the sky, consisted of a small precipice of columnar hyperite, broken by many steep gullies. We climbed a long débris slope to one of these at the west side near the fjord end, and scrambled up it, not without difficulty, for it was very steep, and all the rocks were rotten, so that neither hand nor foot hold was secure. The final step was just touch and go.

This difficulty passed, there came the great rolling upland we had expected, almost flat over areas big enough for ten cricket-fields. It undulated away in all directions, and easily led us on to widening views and new prospects. Now we looked far over the fjord and saw Post Glacier better than ever, and, for the first time, a flat snow col at its head, beyond which was mere sky—not a peak jutting over, nor the sign of a snow dome. Corrie Down, as we called the place, was dotted with a sparse vegetation of the usual kind—poppies, dryas, and saxifrages, all past the prime of their blossoming; but stony areas predominated, and, near the snow-beds, stone-bogs rapidly drying up. Upward we wandered on, following a little brook till we came on to the back of the broad ridge and could see into De Geer Valley and across it to the group of mountains between us and Advent Vale. They were surprisingly fine, and, for me, unkindly complex, a maze of peaks, standing up in apparent confusion one beyond another and pouring down many glaciers, the existence of which would

TEMPLE BAY FROM CORRIE DOWN.

not be suspected from the fjord. Two glaciers almost flowed together at the mouth of a valley opening opposite us out of

De Geer Valley. They spread back into many bays, and were lost in the recesses of the dark hills. Altogether, this view was a revelation. Farther round, over the burnished water and beneath the sun, came the ever clear hills behind Cape Boheman, the hills that enclose the great, gently sloping glacier that spreads back so far and forms so conspicuous an object from Advent Point. The peaked and varied forms of these hills are in marked contrast with that of the levelly bedded type which predominates at the east end of Ice Fjord, whilst their situation makes them visible from afar, and always across a wide stretch of water. They are bathed in perpetual blue, and we often saw them projected against a yellow sky, with grey mottled clouds above a low horizontal line just clear of their tops. For some reason, this year they enjoyed more fine weather than other places not far away. When clouds broke from about us we used often to see this range of hills shining in clear sunlight, like islands of the blest.

Making a wide circle round the head of the corrie, we again reached the hyperite precipice, just above camp. Seen from above, its jutting columns and detached towers stand out in bold denticulation, a splendid foreground to the blue fjord below. This time the cliff was passed by an easy gully, and the talus below descended with a facility the exact counterpart of the labour of ascent. The trend of slopes directed our steps to the bed of the corrie, about half-way up it, where we encountered what is probably one of the smallest glaciers in the world—a little white thing, not more than a quarter of a mile long, nor over thirty yards wide. It was much crevassed at its snout, and was making an infantile effort to advance by throwing forward its little seracs. In reality it was in its second childhood, for once it was quite a fine glacier, with a névé on the downs, a great ice-fall over the hyperite precipice, and a snout down by the fjord, where we saw the still existing moraine. Now

ADVENT BAY HILLS SEEN FROM CORRIE DOWN.

it is merely fed by avalanches from the sides of the corrie. At one point this little glacier has to pass over a lower bed of hyperite, some two feet thick, which protrudes from the softer masses of the hill-side. Over this bed it goes in a ridiculous little ice-fall, almost pathetic in its mimicry of the grand cascades of ice that dignify its greater contemporaries.

We arrived in camp thoroughly fatigued. The fact may as well be put on record. A four hours' walk had been quite enough for us—indeed, more than enough. Energy was barely left to cook dinner and crawl into sleeping-bags. In Spitsbergen all suffered in this way, yet we were not a weakly set. The reason was not easy to find. Sometimes we said it was the food, but change of food made no difference. Sometimes we thought it was mere slowness in coming into condition, but the passing weeks and the continual exercise brought no improvement. We were always slack, intellectually as well as physically. It was a labour to write, a labour to settle down to any work whatever. Yet the air seemed brisk, and came either over the snows or the sea. Purer air can scarcely be found. Though pure, however, it was certainly relaxing, and made life laborious. It possessed none of the stimulus of alpine breezes.

There was still no sign of any boat, and now the utmost limit of Trevor-Battye's time was spent. Why had he not come? We started at every sound without, thinking we heard footsteps on the beach, or calls from the sea—all fictions fancy framed. During the night, indeed, visitors came to our camp in search of meat, gulls that pulled the reindeer joints about, and then a fox. Gregory was on the alert, and drove the gulls away, whilst he shot the fox a yard or two from my tent; but I slept on through it all, and did not even hear the report of the rifle fired so close to my head.

It is hard in writing a book of travel to convey to the reader a truthful idea of the stages of suspense, expectation, and disgust through which one passes, without wearying him with repetition of details. Every hour that now went by was filled with nothing but wonder about the missing boat. Why did it not come? Had it gone to the bottom, or been wrecked on the shore of the fjord? Then there was our faithful comrade, Garwood; he must have reached Advent Point full two days ago. Why had he not brought a boat round for us? Was it that he could not get any of the Advent Bay boats? Or had some misfortune overtaken him? Had he missed Williamson and the ponies, or had the ponies gone astray? or had he come to grief on a solitary climb, or fallen into a crevasse? Such were the questions we spent hours in debating, but never an answer did we find, and never a sail hove in sight. The sun went round and round in a brilliant sky, and days possible for work were taken away one by one, whilst we remained helpless by the now well-known shores of Sassen Bay, skinning foxes and cooking, and watching the deserted fjord for a sail that did not appear.

No previous day that we spent in Spitsbergen was more glorious for cloud-effects than this. In the early morning fine masses of cumuli marched across the southern hills, white and tall, their bases high over the mountain crests. In the afternoon a faint mist, delicate and grey, brooded upon the fjord and almost obliterated the feet of the hills to the north, but let their tops appear against the sky in absolute distinctness. Then came low-sweeping beds of soft white cloud, some thousand feet above sea-level, which crawled about like huge dragons, clinging to the hill-sides. One hung on to the Temple. Another crept out of Klaas Billen Bay and round the nearer headland, casting a deep shadow on the water. Meanwhile above Cape Thordsen a wondrous cloud-drama was played, and every variety of mist, densely packed, and rounded, or finely drawn and evanescent, all shining in sunlight, climbed about the hills or draped their slopes, whilst the faint grey haze lay below, and a sky mottled with brilliant flocks of cloudlets canopied the whole. For foreground to this lovely prospect was the crisp water, tumbled into frequent waves by a steady breeze which also blew in from Hyperite Hat, and finally stranded at the very doors of our camp a splendid block of ice, melted into spires and knobs of clearest crystal, wherein the sunlight played, and about which the waves broke in glistening foam. It was like Galatea's chariot, but alas! there was no Galatea, and no mermen and maidens to grace the laughing shore.

Out of sheer idleness I gathered a pile of driftwood together and set it burning, freeing at all events some of its long-imprisoned and oft-frozen particles, to fly into the air and escape, perhaps to warmer climes, more probably to fall on the neighbouring snow and remain in this Arctic prison. Where, I wondered, did the various trees grow, and who was J. N. P., whose initials were carved on a bit of board? Up to the sky they all went in a blue cloud, leaving only a few ashes behind amidst the hyperite blocks that made the hearth. Gregory had fossils to trim and pack, so he was saved from utter boredom, and could sit chipping away in a relatively happy state of mind; but I had nothing that needed to be done, and was in no inventive humour. I could but watch the clouds growing heavier, and note how the sunlight became golden as the sun went to the north, dipping visibly lower than it did a month before, and already beginning to foretell the day of its long setting.

Ultimately the sky was wholly overcast, and the hills buried in cloud, which powdered them over with new snow, so that when we awoke (July 27) the outlook would indeed have been gloomy, but for the brilliant appearance, about two miles off, by the De Geer Valley delta, of a sloop riding at anchor, and two white tents pitched near it on the shore. Glad men were we, and sprightly of movement. Breakfast was cooked and eaten at a rare pace, and off we went along the resounding shore, caring neither for bogs above nor rocky impediments below. The sloop seemed asleep. There was no watch on deck. The tents were tightly closed, and a boat was drawn up alongside. But our halloo was answered from within. They were the tiniest and simplest tents imaginable, pathetically home-made. There was just the ridged roof of thin canvas, supported by crossed sticks at both triangular ends, and the whole held in place without a single rope by a row of stones piled all round its edges upon the ground.

I went down on my knees and pushed my head in at the door. At the far end of the little tent sat a small and kindly-faced man, crouched like an oriental, while a lad, clearly his servant, lay across the opening. They were eating a simple breakfast. Science was written on the worthy man's face. You would know him anywhere for a collector of something. The huge pile of botany paper close at hand showed his particular business.

After salutations we began asking questions, to which replies were given in very deliberate, but good English. "Is that your sloop?" "No; it is a Tromsö sloop, and has come here after reindeer." "Would it take us to Advent Bay?" "I don't think so. The captain does not like Englishmen with guns. There was an Englishman at Advent Bay, who asked to be brought round here, but he would not take him because he had a gun." "But we are not Englishmen with guns; we are scientific students." "Ah! that is different;

The Colosseum Mountain from Ekman Bay

perhaps he will take you, for he is not a bad man. He is a good man. He brought me round here for nothing, and treated me very well." "What is his name?" "He is Captain Ekrem of Tromsö. If you like, I will row out with you to him; but it is quite possible he cannot take you, for he has landed his men, and some have gone up that valley to shoot reindeer, and others have gone up the Sassendal. He is waiting here for them."

We rowed off to the sloop, and the boy climbed on board and awoke the captain, an uncommonly comfortable-looking, well-dressed sailor, about thirty years of age. He put on an ostentatiously indifferent manner when he saw us. Mr. Jörgensen acted as interpreter. "Oh yes," he said, "I could take you to Advent Bay, but I should lose time, and you must pay. How much would you pay?" "Well, tell us how much you want." "I don't know; it is for you to say what you are willing to give." With the breeze that was blowing, the run would have taken perhaps three hours. There was a man and a boy on the sloop. I offered to pay about a guinea (twenty kroner). "He says that is not enough." "Well then, how much does he want?" "He doesn't know. He says in one hour on shore he can shoot a hundred kroners' worth of reindeer, and that he will lose several hours going with you. In fact, I believe he would not take you for less than a hundred kroner." I made again a reflection that has occurred to me both before and since; the farther north you go in Norway, the more shameless is the swindling an Englishman has to encounter. The simple folk within the Arctic Circle can give points to the veriest Shylock. We bade the man go and make his hundred kroner, and left him. He lolled below and returned to sleep, nor did he make any attempt to land during the whole day.

We rowed back to the botanist's camp, to land him and his boy, then rowed away in his boat, which he most kindly placed at our disposal. It was not a light boat, and the sculls were anything but handy. There was a good deal of sea on for Sassen Bay, and the waves caught us on the quarter, so that we lopped and toppled about; but we pulled lustily on, and reached our camp without an "easy." Then came the tug-of-war. Our united strength did not suffice to haul the boat up, and the tide was rising. She had to be left bumping against the shore. Fortunately she was strong. Striking and packing camp was a labour: such quantities of tins full of rocks, and such a general accumulation, besides three tents, various heads and skins, collecting tins full of spirit, and all the apparatus of scientific exploration—in all, about six hundred pounds of stuff. This had to be packed, carried down to the edge of the sea, and made ready for lading. It was two hours' work. At intervals we hauled the boat farther in as the tide rose. Just when the tide began to turn I climbed into the boat and kept her stern off the beach with a pole, whilst Gregory handed the various things to me and I packed them till there was barely room for the two of us to sit. We certainly could not swing to the rowing nor reach well forward, for bundles impeded every motion.

At last we were able to shove off. I never before appreciated the joy that captain and crew of a ship must feel once the cargo is stored, the harbour quitted, and they safe out at sea. Any tossing would have been a relief after the loading. This, doubtless, is the secret of a sailor's love for the sea. It is not the sea itself he likes, but the harbours he hates, the loading and unloading, the getting in and out. Thus one of the great puzzles of my life was solved—how any one can be found to live on the sea!

Fortunately the wind had somewhat abated and the sea gone down, but we had enough to do to pull our heavy load along. Slowly the cliffs went by. Incredibly slowly our camping-place receded into the distance, but it did recede. We hoped to round the delta of De Geer Valley and reach the neighbourhood of Hyperite Hat, but Mr. Jörgensen awaited us on the beach, and said there was too much sea beyond the spit, and he was afraid for his boat. Our tired arms blessed his fears. We pulled ashore, and landed our camp close beside his.

CHAPTER XVII

BACK TO ADVENT POINT

IT is so easy to have one's tents pitched for one, and such a bother to do it for one's self, especially in a wind, when the canvas bulges the wrong way, all the ropes entangle themselves together, and the whole thing falls, just before the main cord is fixed to its peg. The number of pegs seems endless when you have to knock them in yourself, and the fly, if you have one, becomes a burden. Then all your things have to be put in places where they can be found, and by the time that is done there is cooking to be taken in hand, and everything needs seeking and finding. To move our camp this day a distance of three miles involved about eight hours of continuous hard work, but we thought the time and labour well spent. A new prospect was before us, and a new area for work was opened.

The auspicious occasion was celebrated by a dinner, to which Mr. Jörgensen came with his *famulus.* He brought no news from the wide world, but some gossip of Advent Bay. From him we first learned how Dr. Jeaffreson's party came up with the weekly steamer, and landed on July 10 at Advent Bay. Mr. With, the manager of the Vesteraalen Company, took them and their little Norwegian row-boat across in the *Expres* to Cape Thordsen, and left them there in Nordenskjöld's House. Their boat, being insufficiently moored to the shore, was presently washed away, and came drifting back to Advent Bay, where it was found. The party, thus cut off from return, did not have long to wait. By good fortune some men rowed over at this time from Advent Bay to fetch the winterers' stuff, and Dr. Jeaffreson and his companions returned with them. On the 25th, Dr. Jeaffreson and one of his companions went back to Norway, leaving behind their artist, Mr. R. Huyshe Walkey, who, after a long stay at Advent Point, returned to Europe with Baron de Geer.[1]

Meanwhile the heavens had cleared and the evening was brilliant, so that we could taste all the charm of the new point of view. Not that there were many new hills in sight, but the old ones were freshly combined and better seen. This, for instance, is the point where the Temple, with its long side somewhat foreshortened, shows best. Hence too you look straight up all the length of Post Glacier, which composes beside the Temple and draws back its great even length most imposingly. If a man had nothing to do, he might as pleasantly do it in fine weather sitting at this spot as anywhere else in the summer world.

After dinner I wandered off to the point of the delta, where is a raised gravel bar, recently elevated from below the waters. Gregory followed, intending to collect in and about the small lagoon behind the bar. We began wading streams, and so moved on from point to point, till it became obvious that we had gone for a real walk instead of turning into bed. We crossed the whole delta, sometimes in beds of streams, sometimes on gravel spits, sometimes in soft mud, but oftenest over mossy bogs, the mossiest we had anywhere seen. Beyond the delta came the steep hill-side again, with the hyperite cliff apparently faulted some way on.

[1] The story of the misfortune of this party with their boat at Cape Thordsen found its way in an exaggerated form into the London press. It was stated that their "walrus-boat" was "pinched on shore in the land water on July 16 near the mouth of Dickson Bay by the heavy pack-ice," &c. The actual facts, as above related, were confirmed to me by Mr. With, and by ice-master Bottolfsen of the *Expres*.

The fault became our goal, and its existence was duly established.

This hill-side and the marsh are some of the best places in Spitsbergen for flowers. They grew between the stones and on the mud-banks with a vigour surprising in so high a latitude. But there was nothing this evening, save the snowy heights, to suggest that we were within the Arctic Circle. The sun was warm, the fjord quite clear of ice, the slopes green. More than ever was the comparison suggested with a fine spring day in the Italian lakes; only the air was a little denser here, and the sky a paler shade of blue. If a month of such weather could always be foretold, Spitsbergen would become a favourite summer resort, and its shores would produce a crop of hotels. The midnight sun poured a stream of light in at the tent door when we returned to camp, and made it seem an absurdity to go to bed, the more so that a new sail hove in sight and revealed three men in a boat rounding Hyperite Hat, and possibly bound for us. Expectation often disappointed becomes wary. Perhaps it made us too wary this time. The boat actually contained Garwood with Williamson and a sailor. They saw our tents, but Williamson declared they were not ours, and that we were camped close to Post Glacier. The man is a complete fool. In order not to lose the wind they kept away out and sailed on. Had we fired a shot they would probably have come to us and saved a lot of trouble. They had to row back next morning against a gentle breeze, and when they reached Jörgensen Camp we were gone.

For this day (July 28) opened so mildly, with water so calm and sun so bright, that even the botanist was not afraid for his boat, and gave us willing leave to borrow it and make for Advent Bay. Accordingly, whilst Gregory went through his plant collection with our worthy friend, I packed up camp and in due season we rowed away, leaving with our kind host three joints of reindeer and some bottles of beer, enough I should think to keep him going a long time, for he ate less than a bird. The boat was clumsy, deep in the water with its heavy load, but we settled down to work, following a school of white whales, whose long, ugly backs came bulging out of the water one after another in serpentine succession. All the air and surface of the fjord was gay with birds of the usual kinds, but they were so many and so lively—the little auks diving on all sides, the guillemots hurrying about in companies, the fulmar petrels each going about his own business alone. Several fulmars came to look at us and passed close over our heads, perhaps giving a little squeak as they went, the sole note of emotion that this expressionless bird possesses.

Our first work was to round the delta at the mouth of De Geer Valley. Where streams empty into the fjord, great muddy tongues protrude into the blue waters and wag with the ebb and flow of the tide. The muddy water is a thin film on the top of the blue, and the sculls cut eddies through it and revealed the clearness below. Slow was our progress, yet the bank kept moving by and the view changing. We saw the last of Post Glacier and closed it out. Hyperite Hat came nearer. After two hours' rowing we pulled in at the foot of the promontory, and I landed and climbed it for the view. And what a view it was, from the old tumble-down cairn on the top, built one wonders by whom and what for! There was no new feature in sight, save some hills near the head of Klaas Billen Bay, hills of rather interesting form, with snow domes above one or two of them, and glaciers and snow cols between. But there is always novelty in every view, for light and air give rise to constant change, and to-day the light was particularly sparkling, and the calm bay flashed back not lines but tiny points of sunshine from its ripples. The Boheman glaciers were all faint with the brilliancy pouring over them from the sun.

Rowing on, close under the shadowed and overhanging cliff, which the sun presently peeped round and striped with gold, its really small size was forgotten, and the columnar rocks, though only a hundred feet high, gained all the dignity of noble precipices. In places they were red with lichens, elsewhere spattered with patches of white, signs of the nesting-place of many bird-colonies, little auks, guillemots, and puffins. A rifle-shot fired at the rock filled the air with a feathery cloud, but the birds only swung out a few hundred yards over the water, then rounded back, and settled again; nor did they budge when we approached their nests closely from the shore. There were hundreds of young ones, some already able to fly, others mere balls of fluff. We caught a young guillemot after a chase over fallen rocks below the cliff, under one of which he took refuge. From his cave-like retreat he pecked vigorously at his captors, but once held he was as serene and composed as could be.

While we were inspecting the nests, a boat came round the promontory, the way we had come, and made straight for us. It was rowed with a long and swinging stroke, far more suggestive of Eton than Norway. A few moments settled the doubt—it contained Garwood and two sailors, hurrying to catch up with us, for they had found the botanist and heard from him of our departure. Glad men were we at this encounter. We rowed to a narrow beach below the cliff and landed our things, just where a great snow-bank overhung. The sailors were sent off to take the borrowed boat back to the botanist, and return to us overland. I for one was not sorry for the long halt thus involved, for we had a most lovely picnic ground, sheltered by cliffs behind, and shone on by the full glory of a clear afternoon sun. It goes without saying that there were stories to exchange, but they were only briefly told, for

CLIFFS NEAR HYPERITE HAT.

Garwood was bearer of mails from home, letters and newspapers, which we greedily devoured, each of us seated on a convenient rock, with the water lapping at our feet, the boat bumping on the shingle, and the birds fluttering above from their shelf-planted nests. A knob of ice that afterwards fell close to Gregory's head suggested that we might have chosen a safer reading-room.

Long such charming conditions could not be expected to last. The change came very gradually. A white bank of fog formed upon the water, blotting out all save the tips of the hills. It lay almost motionless at first, only drifting or crawling about; presently it condensed into clouds, which rose and strewed themselves about and over the hills, darkening the landscape and chilling the air. This was the end of the Spitsbergen summer. With the turn of the tide came a wind, far too fresh, which soon tumbled the water about and covered the fjord with white caps. When the men came back and we could set forth, wind and tide were against us. The sail could not be used, and hard rowing was our fate for fifteen miles or more. The waves grew bigger, catching our boat on the quarter and splashing over. It was a bucketing row. Gregory and I landed for a time to walk along the shore for survey and collecting. We found a cauliflower and half a pine-apple on the beach; not brought by the Gulf-stream from warmer climes, but by the great Hamburg steamer *Columbia*, which had recently visited Advent Bay with a cargo of tourists. The pleasant walk on shore soon came to an end, for eastward of Advent Bay are a series of cliffs which must be turned by water.

Now there was rocking and misery afloat, and a cold that froze the marrow in one's bones. Happy were the rowers. I sat or rather lay in the stern on a pile of goods, making notes of the coast-line with trembling fingers and no slight tendency to *mal de mer*, foul visitant, not, alas! confined to the hateful ocean. We turned corner after corner, buoyed up with hopes that at each we should be able to set the sail and run for our goal. But always the wind changed with the bending coast, which it so encircled as to be dead in our teeth at every point. At length the bluffs beyond Advent Point came in view and were greeted with a shout. But still there was elbow after elbow to be passed, each with a huge fallen rock at its foot. The view was in the main invariable—always the sloping, crumbling cliff on our left hand, the fringe of rocks below, and then the turbulent sea. At last the cliffs ended, a low point remained to be passed, and we should hoist the sail and slip into and across the bay. But no! as we rounded the point the wind ceased. The waves remained for an hour or two apparently as large as ever, but there was no breath of air; the tide slackened. For form's sake we hoisted the sail, but the rowing had to continue to the bitter end.

An hour long our goal was in sight, clear beneath the low, thick cloud-blanket. Many were the changes that had been made at Advent Point during our five weeks' absence in the interior. The bare, bleak shore, on whose snowy edge we first landed, now carried a spick-and-span little wooden inn, to which finishing touches of paint and carpentry were being applied. There was another camp too, likewise of green canvas tents, not far from ours, the camp of Mr. Walkey. There were also various piles of stores belonging to the inn, to the Swedish party, or to the men at work patching up the wrecked cutter of the winterers.

About 4.30 A.M. (July 29) we landed, and the dreary work of pitching camp with frozen fingers followed. Our heavy baggage was in a chaotic condition, stowed under tarpaulins and ground-sheets, everybody's possessions jostled anyhow with, under, or above every one else's, tins, hay, sacks of oats, cooking things all mixed together. To render cosmic such a chaos was the work of many hours, and could only be commenced after the tents were pitched, and each became a centre for the accumulation of its owner's particular baggage. The whole of the day that followed after a long sleep was occupied with the same

ABANDONED WINTERERS' HUT.

business, and not till late could we look round in peace, knowing where we were and what we had.

The hours of sleep were hours of heavy rain, which drenched everything we had left about; but when breakfast-time came, perhaps at four o'clock in the afternoon, rain ceased and the weather manifested a dignified reserve. Clouds withdrew aloft and gathered themselves into fine lines and mottled areas, all grey and still, for there was not a breath of air, and the fjord was glassy and leaden, the only ripples upon it being caused by diving birds. Bleared smudges of light showed the whereabouts of the sun. Hills were of purple and pallid white. A great solemnity brooded over the land. Bars of low grey mist trailed across the slopes, constantly changing in place and form, whilst patches of very thin fog gathered on some of the crags like flecks of eider-down. Notwithstanding the clouds, we could see up Advent Vale to Bunting Bluff, Fox Peak, and the slopes and side valleys where we toiled a month before. It was pleasant thus to encounter them again, now that we knew what was round each corner, and could feel that the mysteries and problems they so recently presented had been for the most part successfully solved.

Just as the preliminary call to dinner was being shouted (2.30 A.M., July 30) a boat was seen approaching. It proved to be ours with Trevor-Battye, Ted, and Pedersen returning from Dickson Bay. Great were the rejoicings over the festive board, where all five members of our party were reunited once more. What tales there were to exchange! Each had seen finer glaciers, grander mountains, nobler and more interesting scenery generally than the other. It was Dickson Sound against Agardh Bay.

Long the talk lasted, and many were the mutual explanations. Trevor-Battye had not come to Sassen Bay for us, because he believed he had arranged with the men working at the winterers' wreck to come and fetch us on a given date. They made no attempt to do so, and declared they never undertook to come. There was evidently a misunderstanding, not worth the expenditure of time to unravel. Lost time could not be regained. By hook or by crook our party was reunited. That was the main thing.

CHAPTER XVIII

REPORT UPON EKMAN BAY AND DICKSON BAY

BY AUBYN TREVOR-BATTYE

THE fjord system that penetrates Spitsbergen from the south northward is separable into two principal bodies of water, Ice Fjord and North Fjord. Of these North Fjord, lying north of a line joining Cape Boheman with Cape Thordsen, separates to the north into two divisions—Ekman Bay and Dickson Bay.

Dickson Bay is seldom entered, because its mouth is almost closed by a jut of land; the bay itself is very shallow, and the ice, drifted and accumulated in the narrow entry, may render the exit of a boat impossible for weeks together at a time. At the northern end of Dickson Bay is a neck of land, on the other side of which lies Wijde Bay. This neck connects the two land areas of West Spitsbergen, which may therefore be compared to a (very irregular) dumb-bell. The link obviously offers a possible way of escape to the crews of vessels entrapped by the descending ice-pack in the neighbourhood of the Seven Islands, or of Verlegen Hook, and it would be impossible to overrate the value of a crossing well laid down.

I asked Conway's leave to visit this bay while he was occupied with others of his party in crossing to Agardh Bay. This leave he most kindly gave. His cousin, H. E. Conway, the artist to the expedition, was also to come; at least until

HEAD OF WIJDE BAY.

we found a place for an artist's studies. Conway also kindly spared us Pedersen as boatman, so that we were a party of three. We were to take the whale-boat, and we had a fortnight's leave.[1]

[1] A small misunderstanding here.—W. M. C.

We left Advent Bay on the evening of July 13. It was exceedingly unpleasant with rain and wind, but was clearer over Sassen Bay, into which we had a fine view as we passed. Then the fog fell and we were among the ice.

We had had some idea of landing at the entry by Nordenskjöld's House, but this was impossible. Cape Thordsen itself was almost invisible when we passed it.

July 14. Wind SW. Bar. 29.50.—We had a great deal of ice to work through, but landed early on this morning at the mouth of the Saurie River, between Cape Thordsen and Middle Hook. Here the river and the sea have built up a bar and a beach, behind which lies a little lake forming a harbour capable of holding many boats. We camped on the beach in driving rain with ice-floes driven by the wind on shore, and lifted by the rising tide in places on to the top of the beach and close to the tent.

I went out in the boat again, and landed on a large ice-floe, which might easily have been taken for an island, it was so piled up with stones and mud. I walked along the hills for several hours, but the fog was such that it was impossible to get an extended view. I, however, gained the impression that there is but one valley here, though this of considerable extent. On the seaward side of the mountains to the north, a wide flat of turf and lichen runs for a long distance. Conway with admirable perseverance managed to make a sketch in oils in spite of the wet.

July 15. Wind SW. Bar. 29.30.—Proceeding on our way we pulled for three hours in pelting rain and hampered by ice till we sighted Cape Wijk. Entry into Dickson Bay was impossible; the way was absolutely blocked under the south-westerly gale, and the whole sea to the north was closely packed with floes that ground together and would have stove in the boat. To the west, however, the sea was clear, and we hoisted sail and ran for Cape Waern.[1] We landed on one of its two islands, where many eiders were breeding; and afterwards ran the boat at Cape Waern on to a terrace worn by the sea in the chert rock.

Coincident with our landing fell a remarkable change in the weather—the rain stopped, the wind died away, and we were in sunshine. The impression made both upon Conway and myself by the place and the circumstances of the time, I suppose neither is likely to forget. The sudden clearness, the deep water (literally clear as a crystal), the marvellous wealth of sea-weed, the splendour of the great glacier at the head of the bay, the sharp blue peaks of the nunatak-like mountains which separated glacier from glacier westward over the bay to Cape Boheman; the contrast, on the other hand, where over Cape Thordsen and Advent Bay wreaths of storm-torn clouds were still coiling, and finally a carpet of flowers—andromeda, saxifrages, dryas—more luxuriant by far than I have elsewhere seen them in the Arctics, all contributed to the same effect. However, this report is not the place in which to dwell upon impressions, but just as in the course of our first walk upon the cliffs, we found ourselves speaking in superlatives of the "splendid" glacier and of our plateau as the "flower-garden," so these became the names the two things took.

By the "Flower-Garden" I shall mean the whole of a plateau which is included south and north by Cape Waern and by the range of mountains ending in the Colosseum, and east and west by the sea. The strata of which this is com-

[1] Cape Waern is the name of the point between Dickson and Ekman bays, wrongly named Cape Wijk on the Admiralty chart. Cape Wijk is the point E. of the entrance to Dickson Bay.—W. M. C.

posed dip from north-east to south-west, at such an angle that the chert bed on which our boat lay at Cape Waern was once continuous with the same formation which now makes some of the highest bedding in the Capitolium Mountain, and possibly with those of the east side of Dickson Bay. Going northward, the gradually rising plateau becomes overlaid with carboniferous rock, with here and there dolomitic traces. For large areas over the plateau the chert is so entirely disintegrated, and is laid so level, that it presents a surface like that of a road freshly metalled with finely broken flints. Upon the plateau, separated from one another by a wide interval, are two very remarkable boulders, evidently glacier-borne. They would measure some eight feet in height by eight in diameter, and are schistic in formation. There are several reindeer here, and they are very tame.

July 16. Wind SE. Bar. 29.60.—The wind rose again, but without rain. Conway sketched while I visited the Splendid Glacier, sailing down easily, and landing at the western shore almost at the glacier foot. There is a wide stretch of flat vegetation between the shore and the hills to the west, and I could find no evidence that the glacier was advancing upon it. The current of this glacier is now setting in an easterly direction, as I shall presently show. A result of this changed direction (if the direction, as I feel convinced, has changed) is that that which was once a terminal moraine is now a lateral moraine, if considered in relation to the main stream of the glacier; and evidence is not wanting to show that the old moraine is now partly overlaid by the thin lip of the glacier's edge, and that the débris which falls to the front is actually falling *sideways* as the glacier moves. This western side of the glacier over which I walked for some distance is not broken, nor much crevassed, but is a system of smooth undulations and of glacier lakes.

On the western sea-front of the glacier, and some 200 yards removed from it, is a small island on which I landed. This island is part of a moraine. It was evidently in a state of movement, but from what agency was not so clear, though it seemed to be subject to some pushing pressure below the water.

Upon our Admiralty chart a large island is marked to the north of Ekman Bay, but for this I looked in vain, for reasons which shall presently appear.

We had a very long and hard pull back with the heavy boat in the teeth of the wind.

July 17. Wind SE. Bar. 29.50.—During the whole of this day we had a strong wind with a cold driving rain, which made it difficult to do very much, though I collected many plants and geological specimens. I also walked round the eastern shore of the Flower-Garden, and inspected the entrance to Dickson Bay, which was clearer of ice than before.

July 18. Wind SE. Bar. 29.15.—Nothing could well be worse than the weather. It rained and blew a gale all day and was very cold.

July 19. Wind E. to S. Bar. 29.30.—The weather had cleared; it was sunny and still, and for the first time since our arrival we could see Cape Boheman and the other points and glaciers just across the bay.

I was under way early in the morning to explore Dickson Bay, leaving Conway with firewood and provisions for a week, and taking Pedersen with me in the whale-boat. After four hours of working through the ice-floes we came into clear water, and a breeze springing up, we set sail. The entrance to the bay and its lower end were clear of ice.

I landed first at the promontory called Cape Smith. This promontory is fringed by a high beach considerably above the level of the adjacent land. It was thick in driftwood under which I found a vast number of Collembola, which Sir John Lubbock, who has written a well-known monograph on this order, is kindly working out. The Collembola are wingless insects, and many of them are characterised by a complete absence of tracheæ.

Next I landed at "Velvet Lawn," the name I gave to a large stretch of sloping ground covered with an extraordinary dark-green growth of *Dryas octopetala*, which had evidently flowered very early. I could not find a single flower of this plant that had not gone to seed. Velvet Lawn culminates in a strange semicircle of separate mountain bluffs. I called them the "Conclave," they looked so much like a sitting of old gods. Their protection to the north, and the influence of the eastern and southern sun, might account for the early flowering of the plant.

From here northward to the beginning of the tidal mud, the water was covered with "bay ice," as the walrus hunters call it. This is ice which, formed in bays, is still unbroken up, though melted very thin by the early summer heat. Hundreds of seals, old and young, were lying upon this by the side of their old blow-holes. Between this and the shore, however, was a stretch of land-water filled with moving and grounded floes, through which we worked the boat. We landed next at "The Glen," as I named a wide entry among the hills which leads on the north-west to a long winding valley up which Pedersen walked to look for reindeer, but without success. This area is partly a river-bed, partly a stretch of vegetation, and contains two small-sized lakes. To the north the mountains approach the bay at right angles, but southward more gradually, so that the open area tapers in this direction till the mountains again almost reach the bay some two miles lower down. The river from the valley curls as it enters the glen, winding close round the base of the hill which forms the glen's northern boundary, to spread out into a little delta as it reaches the bay. The western side of the glen at its widest part is occupied by a yellowish-grey carboniferous limestone hill, overlying a base of red Devonian, very curiously hollowed out, and resembling a gigantic arm-chair, so that I called it the "Giant's Chair."

From the seat of this chair, which was partly filled with snow, two large avalanches of stones poured down with a loud noise.

July 20. Wind SW. Bar. 30.50. Ther. 72° Fahr. in sun at ten A.M.—Piled on the beach of the glen was very much drifted wood, and we were able to build with this and the sail of the boat a place to sleep in.

The morning broke clear and sunny. I climbed to near the summit of the Red Hill, 1100 feet by aneroid, and inspected the head of the bay. Red Hill is the first mountain on the western side of the bay, entirely of red Devonian. Thence to the north all are Devonian, or only the highest peaks are capped with carboniferous rock.

Not only is this red colour in itself most striking, but the purples played upon it by the sun are intense, and give a peculiar beauty to the whole head of the bay.

In trying to go northwards from this point we narrowly escaped running aground. Thereafter we touched ground several times, and had much difficulty in getting off. We could not punt, because the floor of the bay was of the softest boulder-clay, so we pulled through the mud inch by inch. A very considerable portion of even the upper, navigable part of the bay is exceedingly shallow, up to the point where the bottom suddenly rises by an escarpment some two feet high, and becomes a tidal mud-flat quite exposed at low water. We eventually worked our way across, and camped at the point I have marked in the map as Dome-View Camp.

July 21. Wind very changeable, N. to S. Bar. 29.55. Clear.—I called this Dome-View Camp because of a striking dome of snow, which is visible from this point, standing up against the blue sky to the north of the bay.

I began the day by walking towards the head of the bay and back, about eight miles in all. From the point I reached I had a clear view of the moraine which masks the north-west glacier, and determined to follow that route later on. I did not know that it was a glacier; it does not so appear, but looks like a valley filled with the moraine of an old glacier now extinct, so entirely is it masked by the piles of débris at its foot. About mid-day I left again with Pedersen, who was to come for companionship as far as the head of the bay, and then carry my gun back. The distance from our camp to the foot of the moraine was about eight miles. We did not accomplish it without some difficulty, due to the river-torrents we had to wade; and when we foolishly tried to cross the mud at the head of the bay we sank so deeply, or rather were so sucked in, that we ran considerable danger of remaining there altogether. This moraine is very extensive; it is formed of piles of sand, and clay, and talus, and where a small stream has cut its way through the centre of the moraine, the walls rise on either side in places to a height of thirty feet.

This, however, is not the main stream of the glacier, which is more formidable altogether, and passing under the hills to the north of the glacier, breaks up into many fingers, and forms, with the streams that enter from north and north-east, a wide shelving stony track at the head of the bay mud. We reached the face of the glacier itself at the end of an hour. This face presents three smooth, black, convex bastions, and arises precipitously to a height of perhaps seventy feet. It would have been unscalable, but for the fact that at one point, and one only, a talus slope led to the top. The glacier itself for some distance was covered with large conical heaps of mud with stone tables supported on pedestals of ice, and with liquid hollows of clay. It is a large and wide ice-field, and is the confluence of various glaciers. Many glacier-streams find their way over this ice-field. Sometimes they are but shallow troughs; sometimes they cut down into its depth in blue ravines; sometimes they are suddenly lost, falling perpendicularly into bottomless caverns, or reverberating from glacier mills. But none of this ice-field is, strictly speaking, crevassed. I made a careful observation with the compass bearings upon position, rise, or incline of the various glaciers and cols. It was cold work, however, especially as we had become very wet in our passage of the rivers and the bay, and it was not long before I sent Pedersen back and travelled on alone.

The glacier marked at the north-east of the ice-field was the one I determined to attempt, because it was clear that its direction promised best. I should at least, I hoped, be able to reach some point from which I could see down to Wijde Bay. In this I was disappointed, as indeed I deserved to be, without ice-axe, companions, or ropes upon a dangerous glacier. This glacier, unlike all the others which flow into the ice-field, does not meet it by an insensible gradation, but drops suddenly, and presents a snout which has a rounded backbone split by a deep crevasse at every few feet. At first these were narrow, and I had no difficulty in stepping across them, but they got worse and worse, until having crossed one by a snow-bridge which yielded under me, I was brought up by another of considerable width which offered a crossing at one point only, where an immense wedge of ice appeared to be firmly jammed between its walls. Having no ice-axe with me, I was obliged to test this very cautiously with my toe, when the whole thing fell in powder and crash. Hitherto the crevasses had been plain to see, for the snow was almost all melted away; but even could I have passed this crevasse I could have gone little farther, for between the point and the col the glacier was covered with snow.

Although I was unable to reach it I could see the col against the sky-line a mile or so farther on, and I have no doubt that this will some day prove the passage to Wijde Bay. By the time I returned to camp I was ready to sleep, as I had walked for twenty-one hours that day.

July 22. Wind NW., cold and sunless.—On this day we left Dome-View Camp at 9.30 P.M. for the Flower-Garden. We stopped by the way to take sketches of the bay under Lyktan, where we could not land because of the shallows, though we landed again at a point lower down, as marked on the map. Here lies a clear little bay and a deep water. I will now describe the leading features of the eastern side of Dickson Bay.

Owing to the peculiar form of the weathered limestone rock which overlies the Devonian, the tops of the hills on both sides of the bay bear fantastic resemblances to castles and crowns, and other objects. Thus Lyktan was so named by Stjernspetz from its resemblance to the lantern of a light-house. In the valley which lies about its base there is a central mountain resembling a Moorish citadel, and another not unlike the Taage. From this valley three dales branch, one to the north-east, one to the east, and a third to the south-east. The entrance to the third is from the bay only just visible where it leaves the second of these three.

The bedding of these mountains inclines towards the south-west, the red Devonian being gradually lost until in the long Table Mountain marked on the plan it appears only as buttresses about its base. Again towards the south the grey limestone is overlaid with seams of chert, and these again by yellow rock of dolomitic character. The spit which so nearly closes the entrance to Dickson Bay appeared to me, although I did not land on it, to be a carboniferous terrace denuded to that point.

July 23. Wind N. to SE., clear, but no sun until night.—After we reached the Flower-Garden at 4.50 A.M. I had so much occupation with writing and specimens that I did little else all day ; Conway during my absence had made some most charming sketches.

July 24.—We now moved to Glacier Camp, Conway and Pedersen pulling in the boat, and I walking down to meet them a distance of about nine miles. We spent four days in an examination of the head of Ekman Bay before we finally left.

July 24. Wind NW. Bar. 29.50.—Flies and mosquitoes.

July 25. Wind NW. Bar. 29.60. Ther. 82° Fahr. in the sun.—Mosquitoes bad.

July 26. Wind NE. Bar. 29.65. Ther. 72° in the sun.—Mosquitoes bad.

July 27. Wind NE. Bar. 29.70.—Blew half a gale and snowed.

Early in this exploration I was able to solve the mystery of the undiscoverable island. The Splendid Glacier is advancing at a rapid rate. It now presents three fronts to the sea, a south-western, a south-eastern, and an eastern front. From these two latter faces the glacier rises in a jagged area of seracs. Between the south-western and south-eastern faces an apex juts boldly out into the sea, and at the time of our visit two immense pinnacles reared themselves from the water, all but separated to their base from the main mass. This double face of seracs, pushed from behind and undermined by the waves, is constantly falling, so that approach in a boat would be a dangerous experiment. The whole of the sea west of the large island, shown by the survey to have existed there not twenty years ago, is now filled by the advancing glacier. More than half of the island is also overspread. Without moraine, without dirt or discoloration, the glacier is pouring over it, and great seracs lie there, separated only, or barely separated, from the flowers and the grasses by the clear stream their drip has formed. A phenomenon more striking than the contrast of the green island with the icy boulders strewn along it, and the grim whiteness which rises so suddenly behind, would be hard to conceive. Nor is this all; for northwards also of the island the glacier is advancing, and at its present obvious rate of progress it cannot be many years before, united with the glacier which comes from the north, it will have entirely obliterated the head of the bay.

Further, since this survey was made, owing no doubt to the resulting change in the set of the tidal and river current, a stony beach of very considerable extent has raised itself from the water under the Capitolium, and forms two sides of a little harbour, where we were able to ride very peaceably during a time of storm.

At the back of our camp rose cliffs of hyperite entirely disconnected from the main mountain, but forming the highest point of the plateau which drops to the Flower-Garden.

Between Glacier Camp and the eastern angle of the bay are the remains of what was once a hut occupied by Russians. The remains of their utensils, of their glass windows, brick oven, and other belongings, are still scattered about.

This part of the bay is very wrongly entered on the chart. From Glacier Camp the coast line diverges suddenly to the east, forming a distinct angle with the Capitolium and a carboniferous terrace which runs up from the shore line to meet the mountains.

I should be sorry to end this report without once more expressing my sense of the kind willingness with which the leader of the Spitsbergen expedition allowed me thus to do a little bit of personal exploration. Although I was naturally disappointed at my failure to cross to Wijde Bay, my account may, I hope, be useful to some future exploring party; for it is too dangerous a glacier to be attempted single-handed.

Not a little of the memory of a very happy time I owe to the unfailing cheeriness and good fellowship, under often dull conditions, of our artist H. E. Conway, my friend and companion at the Flower-Garden.

CHAPTER XIX

AT ADVENT POINT

ONE important piece of work had, if possible, to be done at Advent Point. It was to take observations for latitude and a true bearing of Bunting Bluff. There are discrepancies in the latitude of Advent Bay as set down in the published charts. These will be set at rest by Baron De Geer's observations, which, however, will not be reduced before this volume sees the light. It was highly desirable, therefore, that I should obtain an independent observation. Unfortunately the fine weather was now at an end. Days passed and the sun did not appear. The same clouded sky, which condemned the theodolite to idleness when we were at Advent Point before, reigned throughout the whole of this second visit. Each day seemed colder, cloudier, and more wet than the one that went before. My tent unfortunately had no window. If I closed the doors for warmth it was dark, and I could not be employed; if I opened them for light, the cold prevented any continuous sedentary occupation.

So unpromising, indeed, were the meteorological conditions, that I should have quitted the place at once if it had been possible so to do; but it was not possible. Of our four sledges, two were wrecked and two were in hospital at Hammerfest. They ought to have been returned a fortnight before, or two native sledges ought to have come up in their stead, but the north Norwegian takes his time over everything. In fact, apologies only, not sledges, had come.

It was promised that they should arrive by the tourist steamer *Lofoten* on July 31. There was nothing for it, therefore, but to wait and make preparations as complete as possible for a fresh adventure inland. Thus two whole days slipped away in chilly inaction, the cloud-pall growing ever denser and lower, the showers more frequent, and the winds more keen.

The best published series of observations of Spitsbergen meteorology are those made at Cape Thordsen, between August 15, 1882, and August 23, 1883, by a well-equipped Swedish party.[1] They showed that, during the period in question, the average number of hours when the sun shone was as follows: During May, 10.8 hours; during June, 6.7; during July, 7; during August, 6.7; during September, 2.2. August was the calmest month, E., ESE., and WSW. winds being commonest. The average velocity of the winds that blew was moderate and almost uniform throughout May, June, and July; then it rose steadily till November. The quantity of clouds reached a maximum at the beginning of June, a minimum at the end of July, and a second maximum at the beginning of September. Precipitation was slight and diminishing in May, less but increasing in June, still slight but a little more rapidly increasing in July, still more rapidly increasing in August to a maximum at the beginning of September. The mean barometric pressure was very uniform in May, June, and July, but fell steadily throughout August to a minimum at the beginning of September. The mean air-temperature rose continuously from $-5°$ Cent. in May to about $+5°$, at which it stood throughout July; it fell steadily to $-1°$ at

[1] *Exploration internationale des Régions polaires*, 1882–83; *Observations faites au Cap Thordsen*, &c., *publiées par l'Académie royal des Sciences de Suède*, tome i., Stockholm, 1891, 4to. See also *Observations météorologiques de l'Expédition arctique Suédoise*, 1872–73, *rédigées par A. Wijkander*, in the *Mém. de l'Acad. R. des Sciences de Suède*, tome xii., No. 7.

the end of August, and thereafter less rapidly to the first winter minimum in December. From these observations it would appear that July was the finest month, August much less fine, and September still worse. Our experience during the time of our visit was fully in harmony with these conclusions.

July 1896 was, on the whole, a fine month. There were many bad days, but there were likewise many fine ones. The bad weather set in on July 28, and continued thenceforward. I was destined not to see the sun from Advent Point. It is unfortunate for Spitsbergen that its best harbour should be so uninterestingly situated as Advent Bay is. The island is rich in fine scenery, full of novelty to a traveller from temperate regions, full of interest too for every student of nature. But Advent Bay does not command views specially remarkable or characteristic, nor is Advent Point a comfortable spot. It is exposed to every wind that blows, and there is always a draught over it between the valley and the fjord. The ground is damp, the near hills uninteresting, whilst the way to the interior is the worst up any valley we saw. The tourist who spends a single day of bad weather at Advent Point sees nothing not equally visible in foggy and wet hours near the mouth of the Thames.

The moment the *Lofoten* came in and the sledges were delivered, we began to load them up for a start. Our plan was to divide into two parties. My cousin and I were to go inland again with Pedersen and make our way over the hills to Coles Bay, thence to Low Sound, up the valley of the Shallow River, and so back by Bolter Pass. The others were to take the boat and spend the time collecting fossils at Green Harbour and Klaas Billen Bay, classical sites where Nordenskjöld and others found famous things. All were to meet again at Advent Bay on the 15th and sail home together. It was a good enough plan, but the pouring skies damped any enthusiasm there might have been for it. We worked at our preparations in silence with shivering limbs.

Presently the little twelve-ton steamer *Expres* also came to anchor in the bay. She, like the *Lofoten*, was run by the Vesteraalen Company for the convenience of tourists. The

THE "EXPRES" IN ADVENT BAY.

arrangements were these: Each week the *Lofoten* left Hammerfest with a complement of passengers, showed them Bear Island and the west coast of South Spitsbergen, and brought them to Advent Bay. Those who desired might disembark and make the inn their headquarters for a week. During this time the *Expres* gave them an opportunity of running up to see Andrée's balloon at Danes Island, or making some other excursion. On this occasion she had taken a bolder trip, and steamed on from Danes Island along the north coast and down Hinloopen Strait nearly to Cape Torell—an astonishing record, as any one acquainted with Spitsbergen literature will perceive. As a rule, the ice-pack permits no such casual rambling. Lamont, who spent several summers hunting in Spitsbergen waters, with all the advantages of an excellently-found vessel, was never able to get beyond the mouth of Liefde Bay, or, in the other direction, beyond the Rijk Yse Islands. The summer of 1896 was, in fact, a very exceptional one. Quite early in the year vessels returned from the north with wonderful tales of open sea. Captain Bade's excursion steamer, the *Erling Jarl*, advanced (about July 28) to 81° 40′ north latitude before meeting the pack. Away to the north-east the sea was reported clear of ice. The same story was told by others. The *Expres* saw no ice anywhere. From Cape Torell none was in sight. The sea was open.

It is interesting to note what were Mr. Jackson's simultaneous experiences in Franz Josef Land.

"Since the middle of February 1896," he writes, "we have had weather I believe to be unprecedented in this latitude. The thermometers during this time have been hovering close up in the neighbourhood of freezing-point, and we have had in consequence great quantities of snow, which lies deep and wet on the floes. On March 5 the air was as balmy as June. There have been no cold winds to harden the snow on the floes, and it is almost impossible to drag the sledges across it. Our furs get dripping wet, and when the inevitable fall in temperature comes, they are frozen hard as steel." This unusual condition of weather caused a great body of open water to form in the northern parts of the Franz Josef Land Archipelago.

We did not know these facts, but we knew enough to cause a great disturbance in our minds. There endured

DICKSON SOUND

within me a keen desire to continue the exploration of the interior of Spitsbergen, but in such unpropitious weather what could we hope to accomplish? We might indeed cover some more ground, but with clouds so low we should see little, and survey work would be most unsatisfactory. Moreover, the only piece of country to which the ponies could be taken was obviously altogether similar to what we had already mapped and studied. If we could have carried them across the fjord, or if they could have gone on the ice-sheet, it would have been different, but our limitations were precise.

Against this was a great temptation. We were actually on the spot in a probably unique year—the only year for half a century in which the ice-pack had withdrawn so far north. Ought we not to sacrifice everything and make the most of what might be a great chance? If we could do nothing more, we could at all events make ourselves acquainted with the character of the coast scenery, whereof only the vaguest descriptions exist. We had heard of fine mountains by Hinloopen Strait, and of others down Wijde Bay. What were they like? We knew enough of the interior now to be able to interpret inland views in a manner beyond the power of previous voyagers. Thus we could not fail to learn facts of value. Besides, there was the wide area of possibility stretching broad and far before the eyes of hope and imagination. The name of Gillis Land was spoken. It was enough.

I hailed Bottolfsen, ice-master of the *Expres*, and found him willing to entertain my proposal to hire his little boat for a fortnight or so, but he would not be free till he had taken some of the *Lofoten's* passengers to see Andrée, and brought them back—the work of two days. Then he would be at our service if the *Lofoten* or the *Virgo* would sell us coals, and provided his colleague the skipper was willing to undertake the engagement. I interviewed the skipper and the bargain was struck. The *Lofoten* promised to leave coals, and I returned to camp. Joy and gladness reigned there at the tidings I brought. New scenes ahead, new experiences, a wider world, no ponies to drive, no bogs to wade, no sledges to lift over obstacles, no more daily packing and unpacking, pitching and striking, but a tight little boat all to ourselves and the wide unknowable north ahead. At the worst we should see the historic sites of the whale-fishery—the Seven Icebergs, Amsterdam Island, the Norways, Cloven Cliff, the north coast, Wijde Bay, Hinloopen Strait, North-east Land, perhaps, even the Seven Islands, and then —who could say? Oh wondrous world of the unknown, strange and phantasmal like death itself, hail to thee! Thither, oh thither will we, to win thee to ourselves! There glowed a light in every eye, a new sprightliness animated every limb. Gaily we sat down to supper in our flapping tent, and cared no more for winds or weather, save as they affected the new prospects so suddenly and attractively opening before us.

With such ideas to occupy our minds the hours flew by. A new process of crystallisation was set up in the baggage, whose parts no longer assembled themselves about sledges and row-boat, but took quite different combinations. Baron de Geer and his companion now returned to Advent Bay from hard work at Cape Boheman, and we had pleasant visits from them—visits to us most profitable also, for I was able to acquire a great deal of information from De Geer about former Swedish expeditions in these regions. New plans involved a mass of correspondence, a big mail for Hammerfest, not completed in a moment. Thus the busy hours went whizzing by, and always the cold wind blew, and the cloud-blanket hung low overhead, hiding the hills, and making survey work impossible, so that there were no regrets on that score.

THE GRAVES OF THE WINTERERS AT ADVENT POINT.

When the die was thrown, Bergen cast a shoe. Wonderful that none had been cast sooner! Now it was a matter of no importance, but how troublesome it might have been had it happened on the way to Coles Bay. As events proved, the Coles Bay expedition would have been valueless, for weather continued so bad, and clouds so low, that nothing but a route survey could have been made. The clouds sometimes lifted a little for an hour, and showed the hills all white with deeper new snow than we had before seen, but they only settled down lower and thicker than before, till fog even enveloped the tents, and snow fell upon us. The changeful wind blew this way and that, always seeking the entrance of the tents, pitch them how we might. It was a gruesome time in camp.

Gregory, Garwood, and Ted took advantage of the *Expres's* preliminary trip to Danes Island, to be carried by her as far as Green Harbour, there to collect fossil-plants till we came for them on the northward way.

There was a singular lack of the sense of the passage of time as the days went by in camp, for it was impossible to do anything methodically. Evil weather made packing an intermittent occupation, whilst it was too chilly to adhere long to any sedentary work. Nothing was to be accomplished in the neighbouring hills amid the all-obliterating fog. Trevor-Battye and I drifted asunder in the matter of sleep till we came to playing Box and Cox, the dinner of each being the other's breakfast. Efforts to come round together were hopelessly unsuccessful. Long talks with Pedersen, in our mixed common language, were perhaps the most interesting events of these featureless days, for he is a mine of inaccurate Spitsbergen traditions, which he only divulges under close cross-examination.

One day he refused tea, saying that it was a Russian drink; Norwegians, he said, drink coffee. He constantly opposed Norway to Russia, a reminiscence, as we thought, of the old hunters' rivalry in these parts. We asked about the Russian trappers. He said none had ever been here in his time. He thought the last came about fifty years ago. He had never heard of Starashchin by name, but knew about a certain Isak, who, with his family, lived summer and winter for thirty-six years on Bell Sound, having camps both on the north and south shores. Other Russian encampments he knew of near Cape Starashchin (where the lake is called Russian Lake, and the river Russian River), at Green Harbour by the west side of the entrance, at Coal Haven in King's Bay, at Red Bay, at Liefde Bay on the west side (a big camp), on the Ryss Islands and Hyperite Island in Hinloopen Strait, at Cape Roos, on the Russian Islands in Deevie Bay of Edge Land, at Whales Head in Wybe Jans Water, and on the Dun Islands and neighbouring coast near Horn Sound.

This information was slowly extracted by help of the chart. There were many digressions, and it was hard to keep the good man's attention to the point. One digression was caused by his own name. He calls himself Carl Willumsen Petersen, but we happened to notice "B. Pedersen" embroidered on his socks. "So that's your name, is it?" we asked. "Yes, that's my name." "Pedersen?" "No! Petersen with a '*t*.'" "What does B stand for?" "My Christian name." "Then your name is B. Pedersen?" "My name is Carl Willumsen Petersen." "Then who is B. Pedersen?" "I don't understand." "Were those socks made for you?" "Yes! that's why my name is marked on them." We gave it up, and returned to the Russians.

On the morning of August 2, the *Erling Jarl*, a well-appointed steamer with fifty-two tourists of various nationalities, Germans predominating, cast anchor off the point. She was run by Captain Bade, an old Arctic hand, and the inventor of Spitsbergen as a tourist resort. This year, as already mentioned, her passengers had had the rare good luck to be carried to lat. 81° 40′, without the smallest risk or discomfort. They waited some time at Danes Island, hoping to see Andrée go up, and they visited Liefde Bay. They were now to spend three days in Ice Fjord, and then go south to see the eclipse of the sun from Vadsö. Of course the shore at once became animated with clean and well-dressed people, one of whom came to call, bringing an introduction to Gregory. We entertained him to an ill-cooked feast, I fear. I visited Captain Bade, and heard from his lips the thrilling tale of his experiences on the German Arctic expedition of 1869–70. The ship to which he belonged was crushed in the ice off Greenland, and the crew took to an ice-floe and lived for 237 days on it, drifting down the coast. They built a hut on the floe, but the ice broke across beneath it, so that they lost a great part of their provisions and other supplies. They were reduced to direst extremities and shortest of short commons. At last they were able to take to their boats; they rowed round Cape Farewell, found a Danish ship, and were brought by her to Copenhagen, arriving in Europe the very day that news was received of the battle of Sedan.

Captain Bade was full of interest in our doings and of advice for the future. He blew rather cold on our plans, and prophesied that the ice would now come rapidly down on the north coast, and that we should find it difficult to reach the Seven Islands. The best time had passed. A strong north wind was blowing, and we must be infinitely cautious. "When the ice comes," he said, "it comes with an incredible speed. Remember, on our ice-floe off Greenland, we once drifted through two degrees of latitude in twenty-four hours. The sea is clear to-day, to-morrow it is all ice. Don't venture in an iron boat down Hinloopen Strait.

It is a death-trap. You will probably find it closed at the south end, and when you turn back the ice has come down and corked it at the north. You are shut in for good. It is now August, and the ice must be coming down."

Some of the tourists borrowed the ponies and held a gymkana. Then there was a dance at the inn, kept up to an early hour. After it, the company marched down in military order, with a fiddle and a guitar at their head, and paid our camp a visit. There was cheering and photographing by the pale grey light. A few hours later the ship steamed off for Sassen Bay in the continuing drizzle, carrying the De Geer party with it to map the end of Post Glacier for comparison with previous records of its form. Left alone, we continued labouring at the baggage in intervals between rain and snow. The ponies, and what was left of their hay and oats, were to be sent back by the next *Lofoten* to Tromsö for sale. All baggage not further required was to go with them, for the little *Expres* had room for us only at a tight pinch, and no room at all for aught else save coals. Thus everything had to be sorted, and boxes and bales filled and nailed or sewn up, each as completed being piled on one of two heaps according to its destination.

During this process of packing at the end of a journey, mistakes in equipment become emphatically apparent. From day to day on the route one learns what things needful have been forgotten. At the end one discovers what superfluities have been brought, and still more, what useful things, duly brought, have never been used, because they were not in the right place at the right time. Our chief superfluities were in pony food, unnecessary garments, and snow-shoe apparatus. Spitsbergen, though it affords little food for man, is generous to beasts. We had six trusses of hay and as many sacks of oats that were never required. Snow shoes were too bulky to be carried inland on the sledges. The only way to attack inland ice in Spitsbergen is to climb on to the foot of some glacier that comes down to the sea, and go right up it dragging sledges. For this work ponies are useless, as they do little more than drag their own food for a few days.

We were also over-equipped in the matter of warm clothing. Every one whose advice we took was accustomed to regard all Arctic lands from the seaman's point of view. We never really suffered from cold inland, as we did on the *Raftsund* coming up. Ordinary warm winter clothes are all you need in Spitsbergen. Bulky fur things cannot be carried over bogs and stony areas. This fur apparatus was therefore never unpacked. Even our reindeer-skin sleeping-bags were a mistake, though a comfortable one; they were too big and heavy. Eider-down bags are the things for inland travel. Our tents, though small, should have been as much smaller as it is possible to make them, with canvas flooring instead of mackintosh, and made to close far more tightly than English tent-makers dream of, but the stout canvas employed for them was good and none too thick. Tents for Spitsbergen must be absolutely waterproof. Each man should have a waterproof sheet in which to wrap his bundle. These sheets suffice for flooring.

I take it that the principle of all travellers' equipment is this: The heavy baggage and base camp should be supplied in proper proportions with everything needed for the journey. These things should be packed with intelligence and thoroughly mixed. There should be no boxes full of one kind of thing. Every piece of baggage should contain a well-assorted combination, especially in the case of provisions. Secondly, the light camp should be an epitome of the base camp. It is useless to have a box full of rubber-bands, for instance, at the base, when the single one you need inland is not forthcoming; and so with medicines, tools, and stores of all sorts. Such matters cannot be arranged in a hurry. They need to be thought out leisurely at home before the main packing is put in hand. Nothing should ever be received packed from the maker. All goods should be received at home, separate, so that they can be properly distributed from the outset. Of course a most complete index of the contents of every package should be kept, and this in the case, not only of the main baggage, but also of the light camp-equipment. Infinite time may be saved in camp by ability to lay hands at the moment on the moment's need. No agent or mercantile firm can be trusted with this work. The traveller must do it himself. Finally, on the march, a traveller's pockets or satchel, or, at the least, a satchel carried for him by a porter always at hand, should contain an epitome of the contents of the light camp. There should always be some food accessible, always a simple tool or two, always a bit of string, a rubber-band, a note-book, perhaps (if you are going with sledges) a bit of wire, a nail, a screw, and the like. Zurbriggen was the best companion I ever knew in this respect. His pockets responded to every emergency. Mine, alas! seldom do. If an intending traveller keeps these principles in mind he will find the task of equipment simplified, and his work of preparation rendered more serious; but every minute spent in careful preparation at home, is an hour saved to the explorer at a time when hours may be of incalculable value.

WALDEN ISLAND

CHAPTER XX

ADVENT BAY TO THE SEVEN ISLANDS

AT four A.M. on August 5, we steamed away, on board the little *Expres*, for new lands and experiences of a new order. Melancholy indeed was the land we left behind, with its bleak purple shores, sloping up to hills all white with new-fallen snow from a level of about 500 feet, and roofed with cloud. Bleared gleams of misty sunlight cast an added pallor on patches of the view. We thought the weather showed a tendency to clear, one of the many hopes, destined to disappointment, wherewith all our remaining time in these waters was filled. The *Expres* steamed along by the southern shore of the fjord, where the steep buttressed fronts of the hills were stained in patches on their sloping laps, as by upset paint-pots of vivid green—mossy areas varnished with wet. Running for the Fästningen Rock at the mouth of Green Harbour, where Gregory, Garwood, and Ted were collecting fossils, we passed the Orient Company's steamer *Garonne*, on its way to Advent Bay, with friends on board whom we were sorry to miss.

The wind howled, and rain drove in our faces as we

came to near Green Harbour Camp, and blew the whistle. The screaming thing bored the drums of our ears, but such was the turmoil of the air, that it was inaudible ten minutes' row to windward, where the tent was pitched behind a rock. Only the chance that Garwood was on a look-out revealed our presence. By noon all were on board, tightly packed into the little cabin. It was of a truncated V-shape, ten feet wide at its broadest. Down the midst was a passage cluttered up by a stove, a table, and a washstand. On either side was a narrow bench, and then a shelf on a higher level. A man might lie on this shelf, but he could not turn over, for the ceiling was too close down. Only in an area of about two square yards at the foot of the companion would there have been room to stand erect, but that too was filled with baggage. Everywhere else you bumped your head, even when sitting. We were fairly bruised all over, after a week's knocking about in such narrow and angular quarters. The five of us packed in somehow with all our baggage and bags. Two lay on each narrow bench with heads propped up against the ends, and legs like overthrust strata. The fifth was either on the floor or on a shelf. The tents were roomy palaces compared with this cabin.

In such relative discomfort we headed for the sea, and were glad; for there was a sense of freedom now that bogs and ponies were well left behind, and new scenes were at hand. Once Dead Man's Cape was rounded a new world would open, and all the known hills be wrapped away. Round we went in the tumbling sea. The west front of Spitsbergen began to unroll before us, the sea front of the long mass of hard archæan rocks, which keeps the rotten interior of the island from being swept away by the inroading ocean. A series of splintered ridges, striking inland, here abut upon the sea. Their other ends are presented towards Advent Bay. Broad glaciers flow lazily down between them to either shore. One glacier ends just north of the Dead Man. Farther up comes a second and wider glacier, with a great northern tributary. The wide crescent front of the second glacier loomed out of fog, its edge broken into blue-faced seracs. The ice-cliff, doubtless 100 feet high, produced no impression of altitude, but only of width, and the glaciers of flatness and extension, a whole world removed from the appearance of Alpine glaciers. A mountain mass followed, unbroken by big glaciers as far as St. John's Bay. Low rain-clouds hung above the mist, and rain-besoms swept across in front. The very universe seemed melting away.

We were now hurrying up the narrow sound, named Keerwyk, dividing Prince Charles's Foreland (named after Charles, Prince of Wales, afterwards King Charles I.) from Spitsbergen. The Foreland is a submerged mountain range with submarine banks of débris piled about its foot. The southern extremity, Saddle Point, seems to have been an island at the time of the discovery, but is now connected with the rest by a low flat, a few feet above sea level. Unluckily all the mountains of the Foreland, some of the finest and loftiest in Spitsbergen, were buried in cloud. We could only see the mouths of gloomy valleys, and the bases of massive buttresses, solemn rock-forms dignified in their mystery of cloud-envelopment. Our little boat, plunging into the short seas, or leaping on to them, seemed dancing with life; she had a rollicking way with her, and shook off the water from her back like a duck. Our pace was not more than eight knots, but we had a greater sense of swift movement than I ever before experienced at sea; water, air, and boat all seemed to be flying along. It was strange how little floating ice we met, though in close proximity to so many glaciers ending in the sea. It would seem that a wide serac'd glacier front should be fringed with fallen ice, but it was not so. The channel grew narrower; glaciers came down to the water on both sides, and fog fell upon the sea. We plunged into mere grey nothingness, and so passed the narrowest and very shallow part of the strait. A lift of the curtain displayed purple English Bay, with a steepish little glacier tumbling into it from the north. The heavy grey clouds seemed to walk on the sea, on columnar limbs of falling snow, thick, grey, and heavy, like the clouds. The Sound ends between the bold head of Fair Foreland (Vogelhoek) and the low spit named Quade Hook, where King's and Cross Bays open. They are divided by a fine mountain mass, dark and bold, with many valley-laps in it. Each bay enclosed a separate storm, so that the promontories only could be descried.

On we went, almost due north, now in the open sea, and once more in fog so thick that the land was seldom visible. Only three of the so-called Seven Icebergs were seen. They are low, wide, gently-sloping glaciers of the normal type, some pushing crescent-fronted ice-cliffs into the sea, others with pudding-ends stopping short of the waters on a débris flat. Thin splintered ridges divide one glacier from another, and all, I believe, drain the inland ice-sheet lying between the coast and Liefde Bay. Of the nature of this ice-sheet nothing is known. Looking up Cross Bay, we afterwards learned that deep valleys, separated by rugged ranges, penetrate far inland to the north. They are the orographical continuation of the high land east of Dutch Bay. It may be that these mountains form a backbone dividing the ice into two separate sheets. Large glaciers in Spitsbergen do not necessarily imply the existence of large, or indeed of any, névés. An Alpine traveller finds the realisation of this fact difficult. Seven such snouts in the Alps would prove the existence of a feeding snowfield as large as the whole Bernese Oberland. In Spitsbergen nothing of the kind need be postulated.

Of Magdalen Bay, called the fairest jewel of Arctic scenery, only the ghost loomed forth, robed in white and crowned with needle-pointed splinters of aspiring rock. Had it been more plainly discovered, we should not have greatly cared, for the tumbling of the boat and the jarring of the screw had by this time wrought weariness in all. We were

ONE OF THE SEVEN ICEBERGS.

counting the hours to a respite in smooth water. Presently, round a corner, came the narrow and impressive South Gat—once scene of the great whale fishery and entrance to the haven of the Dutch. A sunken rock in the midst of the way needs careful avoidance, but our skipper had already made this part of the voyage many times this year—conveying tourists to see Andrée's balloon. Full steam ahead therefore we went, disturbing countless flocks of little auks and with infinite fulmars swooping around. Thick fog and a heavy fall of snow brought us to a halt as soon as we were in the smooth water of land-locked Dutch Bay, which is marked on the chart as Smeerenburg, though that name properly belonged not to the haven itself but to the old Dutch whalers' settlement on the low east spit of Amsterdam Island. The true Smeerenburg, or Blubbertown, had but a brief period of prosperity, and has long been in ruins.

The west side of Smeerenburg Bay is formed by Amsterdam and Danes Islands. They are separated by Danes Gat, whilst South Gat divides Danes Island from Spitsbergen. Feeling our way through the fog, over water leaden-smooth, we came at length to the little enclosed bay in the south side of Danes Gat, where Andrée's ship *Virgo* was anchored, close by the balloon-house, on the north shore of Danes Island, at the point where once stood the "Cookery of Harlingen." Here we too cast anchor, for the *Expres* needed coal, and we had good hope that the *Virgo's* captain would kindly supply us from his superfluity. Crossing in the dingey to call on Herr Andreé, we were struck by the intense greenness of the water, rendered all the more emphatic by contrast with a brilliant yellow stain on the rocks by the shore, the result of recent gas-manufacture for the balloon.

Assuredly few places in the world can be more utterly forlorn than this rock-bound bay, frowned upon by bare hills, about whose bases angular débris are deeply piled, nothing in sight but barren islands and splintered glaciers, "with black air accompanied, with damps and dreadful gloom." Snow lay deep down to the very margin of the sea, and a thick snowfall was at that moment taking place. The north winds only "bursting their brazen dungeon, armed with ice, and snow, and hail, and stormy gust, and flaw," avail to fall direct upon this enclosed spot. But draughts at all times eddy round and round, and cause the snow-flakes to dance together in columns like restless ghosts. Here it was that Mr. Arnold Pike built a small wooden house, wherein he passed a winter far from the haunts of men. The house was being utilised by the Swedes, who set up their strangely civilised-looking gas apparatus close alongside. Studley, after leaving us at Advent Bay, spent a whole month in Pike's House, waiting to see the balloon go up. Walkey immortalised him in a monumental outline on the door—an unmistakable likeness!

Mr. J. Stadling of Stockholm, our companion on the *Raftsund*, extended a warm greeting to us on the *Virgo*, and conveyed Herr Andrée's invitation to go over the balloon house with him. A few strokes took the boat to the little landing-stage, where Herr Andrée and the two intended companions of his proposed aerial flight joined us. We were shown how the gas was made, and the long silk pipe meandering amongst the stones to convey it into the balloon. The great distended sphere filled the roofless wooden house and bulged out above. Like all balloons, when seen near at hand, it appeared surprisingly large. It is related of a shy curate, who had sat in absolute silence throughout a dinner at the squire's house, that with the coming of dessert he suddenly remarked, *à propos* of nothing, "The cuckoo is a larger bird than you'd suppose." The same general statement I maintain to be true of balloons. They are all larger than you would suppose. There is here no need to describe Andrée's balloon. It has been described often enough. Interesting as it was to me, with all its compact contrivances, it was far less interesting than Andrée himself.

No one could see him and not be struck by the evident force and capacity of the man. In his presence, the idea that any wavering of intention found place in his mind was inconceivable. Pestered, as he had been for weeks, by inquisitive visitors, he seemed on the defensive, and suspicious of criticism in every question. He had been told that his scheme was in every respect impracticable. "They said I could not set up and inflate my balloon in this place. I have set it up and inflated it. They said it would not hold the

HERR ANDRÉE'S BALLOON HOUSE.

gas for a sufficient length of time without leakage. It has now been inflated for ten days or more, and it does not leak. There were two little needle-point holes only, and those we easily mended. We have considered everything and provided against every accident, and now we are certain that, whenever the right wind blows, we can start without a hitch. Unfortunately since July 15 (this was August 6), there has been no south wind. It is not enough for me that the wind should be from the right quarter. I must have a chance of decent weather, so that we may be able to see something. Now it is getting late. The *Virgo* is not insured after August 20, so that, unless a good wind sets in soon, we must pack up and return to Sweden. It would have been better if we had been able to come up here earlier in the year, but we could not, for the balloon was not made in time. Our work, however, has not been wasted. The experience we have gained will be valuable, when we come back again next year, as I hope to do. We shall come earlier, and shall be able to get the balloon ready quicker. There is always plenty of south wind early in the season. Three days of a moderate wind, blowing approximately from the south, is all we need. After that the wind may blow how it pleases, it cannot help taking us towards some of the land that encircles the polar ocean. We can remain afloat for three weeks, and in that time, with any luck, we ought to be carried down to some habitable country."

Herr Andrée then invited me to climb to the top of the balloon, and see how the snow that was falling heavily was shed off the dome. The ascent was made by a kind of zigzag wooden staircase, forming a buttress to the balloon house. There was little in the nature of banister, and the space between the steps was open. The steps were covered with fresh snow, on which my rubber boots slipped about. I never felt less secure in my life. A slip at the top would have launched us straight into the air, a hundred feet above the ground.

The *Virgo's* captain placed five days' coal at the disposal of the *Expres*. We took it on board in barrels, and piled them along the narrow gangways, forming all the deck there was to the little boat. The weather was thick as ever when the anchor was raised, but this is a land that bad weather dignifies and adorns The strong black ancient rock, broken into points and draped with whitest snow mantles, seemed able to defy alike raging sea and splitting frost. The mountains all around were of the same bold type, but at the foot of those that form Amsterdam Island is a wide low spit of ground jutting out into the bay, and carrying the ruins and graves of what was Smeerenburg. This shore was well adapted for drawing up the carcasses of whales, killed either in the bay itself or in the neighbouring seas. Here the blubber was cut from them and boiled down in one of the "tents," or factories, whose very foundations have long disappeared.

A heavy mist lay low down, and cast leaden shadows upon the smooth water. Only the bases of the hills could be seen. On these the imagination was free to pile whatever mighty towers it pleased. It was easy to fill the scene with high-pooped Dutchmen riding at anchor, whilst the shore was thronged with busy crowds. For many thousands of men and women, in the palmy days of the whale fishery in the seventeenth century, annually resorted here to catch the fish or handle the produce of the fishery. Large glacier fronts protrude into the sea on the east coast of the bay. A blaze of white light, a true ice-blink, gleamed in the mist over the level glacier surface that comes down with almost imperceptible slope from the unexplored inland ice. All around was grey—grey water, grey sky, grey rocks—save for faint blue breaks in the glacier fronts, and one incredibly deep-blue castle of stranded ice, whose colour, like a rich note of music, seemed to throb in and through the soft and tender harmony of grey.

Our first need was to steam due north and find the ice-pack. The last news we had of it was, that about July 28 its south margin was in lat. 81° 40′ N. But a strong north wind had been blowing continuously ever since, and there was unfortunately little doubt that the ice had by now

DRIFT ICE OFF SPITSBERGEN.

come much farther south. Out therefore into the mist and the heaving sea we hastened, past Foul Point and Vogelsang. Of this island, as well as of Cloven Cliff and the Norways, we had a clear view. They resemble, in structure and colouring, the other shores and islands of the neighbourhood. It was their historic interest that affected me. How many Arctic explorers have sighted them—Phipps (with young Nelson), Scoresby, Franklin, Parry, Nordenskjöld, Leigh Smith, Lamont—oftener outposts of disappointment than of surprise! For in many years the ice-pack comes down to these islands and remains fixed against them week after week. How many skippers have in their turn climbed Cloven Cliff, and gazed towards north and east, only to find ice everywhere and no possible "lead"! Thus far, at any rate, the fortune of the year favoured us; there was not a fragment of floating ice in sight.

In five hours' time, however, we ran into it, in latitude 80° 13′, and turned eastward along its edge, which forced us somewhat south at first, then bent away northward, so that after six hours' running we were in latitude 80° 28′, almost north of Verlegen Hook. Here the edge of the pack turned sharply to the south, and compelled a change of course. We ran along it for another hour or more, and then came to drift ice, broken up and scattered over the surface of the sea as far as Verlegen Hook itself. By twisting about, a way through was found into open water again, at the mouth of Hinloopen] Strait, but the pack was not far off, and was unmistakably coming down on Verlegen Hook.

The broken ice-sheet that fills the Polar Sea continually drifts from the north-east on to Spitsbergen. Divided into two parts by North-East-Land, it opens like two jaws, whereof the north usually closes upon Verlegen Hook, the south upon Wiche Land, where, subdividing, a branch goes to plug the south entrance of Hinloopen Strait, whilst the other jams up against Edge Land. Thus Verlegen Hook to the north, and the Ryk Yse Islands off Edge Land to the south, are frequently the limits beyond which vessels cannot pass. It seemed to be a mere question of hours before Verlegen Hook would be infested. The floating ice was already accumulating against the coast, and soon the gap separating the pack from the land would be filled.

Hereupon a difference of opinion arose on board. The skipper was for turning back. He said it was folly to go forward. The ice was coming down and we should be cut off. The *Expres* is an iron boat incapable of resisting the smallest blow from ice in water at freezing temperature. It would be difficult to avoid touching some of the many ice-blocks that dotted the sea in all directions. He wanted neither to lose his ship nor his life. The ice-master, Bottolfsen, on the other hand, was willing to go on. Twenty-four hours, he judged, would elapse before the ice could close on Verlegen Hook, and in twenty-four hours we might run to the Seven Islands, and see how the pack lay in that direction. Perhaps there might be open water, and we might get round North-East-Land. The matter was referred to me. I said "Go on," and on we went, in and out amongst the floating ice, pieces of all sizes,—small blocks we might have hauled on board, and flat floes big enough for a cricket match. Gradually the open water became larger, and in an hour or so we emerged into clear sea once more, and found the edge of the fast ice trending north, fringed by lines of large loose masses with tilted tables and blue mounds of crushed floes. A sail was seen ahead. It proved to be the stout-looking, new-built walrus-sloop, *Lykkenprove* of Tromsö. We came to and hailed her. The warmly-clad crew and cheery captain assembled on deck. When our engines stopped, and we lay still on the now calm water, the silence of the great deep was almost oppressive. Through it the voices of Bottolfsen and the sloop's skipper rang, as might those of challenging heroes beneath the walls of listening Troy. We learned that, a fortnight before, the sloop had been hunting round the NE. corner of North-East-Land. Then the ice came down, and they were forced to run before it. Four days ago they passed the Seven Islands but little in advance of the ice, which by now had certainly reached them. Their catch had been poor. They had killed one bear, and filled ninety barrels with blubber—that was all.

A faint breeze carried them slowly westward, whilst we steamed away north-east near the edge of the pack. The coast of North-East-Land lay about ten miles off on our starboard hand, bleak, desolate, and cloud-capped, in type altogether similar to the north coast of Spitsbergen, and the islands near Smeerenburg. Deep new-fallen snow lay thick on the old snow, down to the sea's margin, only cliffs and steep rocks stood out black from the white mantle. The existence of the great inland ice-sheet, by which this island is wholly enveloped, was here and there suggested through breaks in the clouds. Milton, who owed so much of his knowledge of strange lands to study of *Purchas His Pilgrims*, might have had North-East-Land in mind when he wrote :—

> "Beyond this flood a frozen continent
> Lies dark and wild, beat with perpetual storms,
> Of whirlwind and dire hail, which on firm land
> Thaws not, but gathers heap, and ruin seems
> Of ancient pile ; all else deep snow and ice.
> A gulf profound as that Serbonian bog
> Betwixt Damieta and Mount Casius old,
> Where armies whole have sunk ; the parching air
> Burns frore, and cold performs th' effect of fire.

A universe of death, which God by curse,
Created evil, for evil only good,
Where all life dies, death lives, and nature breeds,
Perverse, all monstrous, all prodigious things,
Abominable, inutterable, and worse
Than fables yet have feigned, or fear conceived,
Gorgons and Hydras, and Chimæras dire."

A few hours later the remote Seven Islands appeared ahead, matching in form the bold bluff of North Cape Island's northern front.

We did not run directly for the islands, but headed to pass them to the west, hoping to steam round and find again the edge of the pack, which had bent away and left open sea all about us. If the north coast of Spitsbergen had seemed bleak, these islands, as we neared them, seemed yet bleaker, yet more desolate and aloof from man. First came Walden, an arête-crested mound of hardest rock, defying the inroads of the sea. Beyond it we saw, as one mass, the larger islands of Parry, Phipps, and Martens—all cliff-sided, bare, and lonely. The reader will find it hard to share the emotions evoked by the sight of these islands in the mind of one to whom, by much reading of books of Arctic travel, they had long become, if inaccessible and remote, yet definite realities, associated with the doings, the struggles, and the disappointments of great explorers, memorable in the annals of daring and human achievement. There they lay silent, cold, and still, under their pall of cloud and snow, with the gloom of the north enshrouding them. Hardly a bird skimmed the surface of the forsaken sea, wherein only ice-blocks floated. To the north-west, beyond the three larger islands, a few lonely rocks stood forth, joining clouds and sea—the two Table Islands and Ross Island—last outposts of land towards the Polar Ocean, which, a few miles farther, sinks to a depth of 1370 fathoms, as Nordenskjöld discovered.

ROCKS OF WALDEN ISLAND.

We were destined not to reach the Table Islands; on approaching, we found them enveloped in the pack. The report of the *Lykkenprove* was true. We were hopelessly cut off from the east. The time had passed in which it would have been possible to advance far towards Giles Land. Accordingly, after again coming as close to the pack as we dared, in lat. 80° 39′, our greatest northing, the boat was put about, and headed for Walden, through a sea clear of ice, and of an incredible peacock-blue colour. This was the island on which the Wellman expedition took refuge, when their steamer, the *Ragnvald Jarl*, was nipped in the ice and destroyed close to its shore on May 28, 1894. Bottolfsen had been ice-master to Mr. Wellman, and could tell many details about the adventures of the ill-fated expedition.

We passed round the north promontory, and came to, off a bay in the east shore. The sea behind the island being very calm, it was possible to land, so we lowered the boats and rowed ashore, glad to feel solid rock beneath our feet. Solid, indeed, are the red and grey rocks of Walden, foundation stones of the earth's first crust.[1] Yet, hard though they be, the violence of Arctic weather avails to crumble their surface into a kind of rough sand, which can readily be scraped off by the boot. There is no soil on the island, save in narrow gullies and chinks protected from wind. In such places snow-beds also permanently linger. The shore is eaten away into little coves, intricately bent. The surface of the rock is scored into deep undulations of ridge and gully, and every yard of progress involves a scramble.

We made our way to the ruined framework of the Well-

[1] Dr. Gregory and Mr. Garwood will describe the geology of the island in another book.

man hut, drawn by the resistless attraction of a human interest. Footprints in a steep gully piled with sand showed the way to it. They had been preserved beneath a covering of snow. Truth to tell, there was little enough to see—a mere framework of beams, the wreck of sleeping-bunks, floors, and doorways, a heap of coal, piles of withered-up potatoes and peas, foul remnants of old clothes, empty cartridges, a packet of photograph developer, and such like rubbish. It was interesting to hear Bottolfsen's reminiscences. "When we opened the door one morning there was a big bear standing close to it, just where you are now. He seemed to be waiting for some one to come out. I caught sight of him in time and called to the skipper to look out. I only stopped him just in time, for the bear would have been on him in a moment. We shot the bear. Perhaps this may have been his skull, though I don't think so—it is not large enough."

Strolling about, looking at the rubbish, I came upon what looked like a candle-end wrapped in paper. It seemed too hard for a candle, and I threw it violently on the rock at my feet, to see if it would break, for it was hard frozen. A yard or two farther on was a pile of similar objects. "What are these?" I asked. "Oh," said Bottolfsen, "those are part of the case of dynamite!" I did not try to break any more of them.

The back of the island's ridge afforded a fine view over sea and land. Clouds had lifted somewhat, and the larger islands more amply displayed the lonely grandeur of their weather-beaten, snow-draped flanks. Even nearer at hand to the south were the crags of North-East-Land's North Cape, with the Castrens Islands by it, whilst far to the east Cape Wrede and Cape Platen lifted their bold fronts beneath the cold, white blink that showed where the fast ice bound the

RUINS OF THE WELLMAN HUT ON WALDEN ISLAND.

sea. It was a memorable view. We gazed at it, as men looking upon a world they will never see again. The little *Expres* lay still upon the calm water in the island's lee. She seemed impatient to be off, fearing the adventure of the ice, which was coming down with stealthy drift, nearer and nearer every hour.

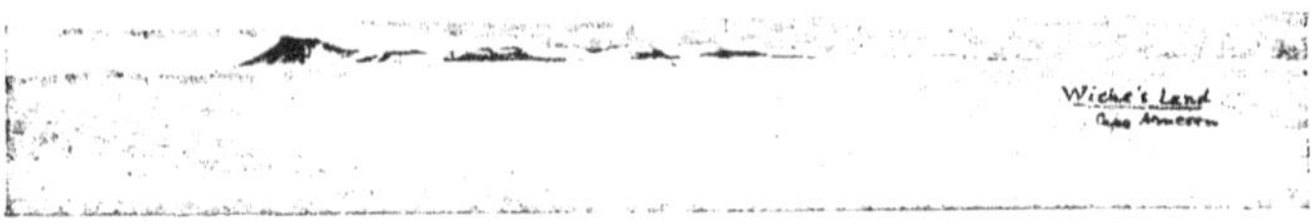

WICHE LAND.

From a Sketch by A. Trevor-Battye.

CHAPTER XXI

HINLOOPEN STRAIT AND WIJDE BAY

AFTER noon (Aug. 7) the engines were again set going with the ship's head to the SW., almost retracing her course. At first there was some movement of waves, but when, in about five hours' running, the loose ice was entered, the surface of the water became deathly calm. Off Verlegen Hook the question of whitherward was again mooted. The skipper was for running west, beyond the ice-trap. We preferred any alternative to that. Hinloopen Strait opened invitingly close at hand. Why not run down it, and attempt to circumnavigate Spitsbergen—a feat never accomplished by travellers. If a way was open to the south, it would not matter whether Verlegen Hook were blocked or not. The same north wind that was bringing the pack down on it, was possibly driving the ice away from Olga Strait. Of course, we might find the south entrance to the strait blocked, and be driven to return, and then find ourselves cut off at Verlegen Hook. Walrus sloops before now have been thus circumstanced, and their crews have had to abandon them and beat a retreat in their boats, not always avoiding tragic issues. Such possibilities added zest to the enterprise. We determined to accept the hazard.

NEAR THE MOUTH OF HINLOOPEN STRAIT.

Round went the rudder and away sped the *Expres* on a new track. In an hour's time we had passed the long low flat of Verlegen Hook and entered the mouth of Hinloopen Strait. The air became warmer and more soft, under a thoroughly English winter sky. Rain-clouds travelled over land and sea, and grey brooms, sweeping along, showed where snow or rain was falling. They were more common in bays and over the land than on the sea. One dense besom swept Treurenburg Bay and hid the point behind which, in Hecla Cove, Parry left his ship before his memorable expedition to the north over the ice-pack. East was North-East-Land; its shores sometimes low, but always with a cliff-fronted plateau behind, swelling up to a vast area of snow that vanished into cloud. Little auks again peopled the waters in considerable numbers, and went flopping away from us, beating the surface with their wings; or dived hurriedly out of sight. The scene on all sides presented an effect of breadth and lowness. Even the cliffs looked very low in contrast with their breadth. White promontories jutted into the water. White hills stood forth against black clouds, and dark rocks against white patches of fog; for the air was full of vapour and wet, and was mottled light and dark according to its density. The sea in the straits was perfectly calm, and the prospect of its so continuing enabled us to come to our right minds. The cabin was put into some approximation to order. Sketches were made, guns cleaned, and slides changed in the cameras. Cooking was put in hand, and general cheerfulness reigned.

From Hecla Hook to Lomme Bay, the whole coast of the part of Spitsbergen called New Friesland is formed by one long glacier front, twenty-three geographical miles in width. In some places rock beds lift the foot of the ice above sea level, but for the most part the glacier terminates in a cliff of ice actually washed by the waves. To describe this as a glacier twenty-three miles wide is to give a false impression. The whole land area between Wijde Bay and Hinloopen Strait is covered with an ice-sheet, which would flow on all sides into the sea were it not for the configuration of the margin of the land. A range of hills encloses it on the west, so that only three ice-tongues, as we shall

EAST SHORE OF WIJDE BAY.

hereafter see, actually reach the waters of Wijde Bay. To the north, the plateau edge similarly obstructs the ice, and only permits one outlet, down which it flows into the head of Treurenburg Bay. On the east side the rock-walls are fewer, and the ice can flow freely down between the rocks of Hecla Hook on the north, and the Lomme Bay hills, and again (not to mention small ice-tongues) south of

Mount Loven. Keeping pretty close to the ice-cliff we had a better opportunity of watching the form and behaviour of a large sheet of inland ice than ever before.

As in the glaciers near Ice Fjord, so here crevasses were numerous, but all blocked by new snow as soon as formed. The drainage is thus kept on the surface. Water flows down channels, which it cuts into deep grooves. These become arched over by later fallen snow, which swiftly turns to ice, so that the wet channel of one year tends to become a dry tunnel the next. The sections of such dry tunnels, made by the terminal ice-cliff, revealed blue grottoes of marvellous beauty. More impressive, however, than these, and even than the great fissured cliff itself with all its azure facets, was the vast, gently curving ice-field, reaching back so far under the heavy grey cloud canopy, which here and there rested upon it, and interrupted its long low sky-line, casting upon its spotless surface an infinite gradation of the tenderest shadows. The whole scene was marvellous beyond all power of words; most memorable was the gravity of the colouring, the dark-green sea, the purple rocks, the blue glacier cliff, the near grey, the remote yellowish snow, and over all the dull leaden-grey of the clouds, combined into a solemn harmony of tone over which brooded the great silence of the north.

The view defiled before us as we passed, and made the enormity of the scale of the ice-phenomena gradually apparent to the senses. With slow movement the cloud-roof kept drifting from the east, very low down, trailing sometimes a skirt of falling snow, which blotted out everything beneath it. Gloomy Lomme Bay showed the bulging cliffs streaked with snow that wall it round and keep back the ice-sheet except from its head, which a land-locked storm hid from us. Wide low Wahlenberg Bay on the other side contrasts with Lomme Bay in character. Its north side is formed by sloping ice, whilst along the south are a row of plateau fronting débris-slopes, like those along the south side of Ice Fjord. This bay was specially interesting to us, as being the place where Nordenskjöld descended after his famous traverse of the inland ice of North-East-Land. Hereabouts we met another hunting-sloop, the *William Barents.* She stood out, a dark and sharply outlined thing, against the hazy grey and white background. We hailed her, but learned nothing from her secretive and suspicious skipper.

Slowly against the powerful tide we forged onward down the strait, passing the low Foster Islands, and approaching the archipelago that studs the south Waygat, over against Cape Torell. The snow again ceased to fall and the clouds withdrew somewhat aloft, and multiplied into varied forms, soft and grey. Calm water, in grey and green undulations, with a line of dark beyond, led to the low glacier fronts of North-East-Land, beyond which the large ice-sheet delicately sloped up into tender mist and blanched ice-blink, with a dark cloud-dragon lying above all. The view broadened down the low-sided strait. There were black islands in front, sharply edged, but all else, near or far away, was utterly soft and vague.

The slow changes and developments of scenery that matched our movement were to me of infinite fascination. Each hour brought some new effect, some fresh delight. The charm lay not in any one view, but in the succession of slight variations on one theme. But the attempt to suggest by mere words the aspect of these effects and mutations, even to a careful reader, without calling upon him for too sustained an attention, or requiring of his fancy too large an exercise, seems almost foredoomed to failure. The attempt, however, must be made, for these scenes have never before been described, and the chance of be-

ICEBERG IN OLGA STRAIT.

holding them, as we did, is so rare that it may be long before the opportunity occurs again.

We passed between Cape Torell and the low black island of Walberg. In contrast of form William I. Island appeared in the south, high and with a round snow top. Its east bluff is named Thumb Point—a name foolishly common in these parts, and in this case inapt. Over against Cape Torell, a big white nameless valley, floored with a glacier flowing between bluff-fronted plateaus, debouches on a wide nameless bay. This valley would afford a good avenue of approach to the interior.

Beyond Cape Torell came the wider sea, which was called Olga Strait, when Wiche's Land was believed to be larger than it actually is, and the water between it and Spitsbergen might be thought of as a narrow sea. Now began excitement as to the position of the ice-pack. Was the way open to the south, or was it blocked? At present there were but few pieces of floating ice, and those small. Only over the land on either side was there any ice-blink. Right and left the broad cold blare of light lay along the lower heavens, but south and north it faded away into yellow, proving that for some distance the sea was clear before us. The low white south coast of North-East-Land trended away under clouds with the inland ice-sheet drooping down to it. The coast was fringed with loose ice closely packed together. Presently larger pieces of floating ice than we had before seen began to dot the water. They were true icebergs — small no doubt in comparison with Greenland icebergs, but much the largest we saw in Spitsbergen waters. They came from the south-east coast of North-East-Land, where the ice-sheet ends in the sea along a front over sixty miles in breadth. Wind, tides, and currents had brought them round Cape Mohn, and scattered them in our path. They were of infinitely varied form, broken into towers, hollowed into caves, heeled over and fallen together, or cloven asunder in every direction. Blue tones shimmered upon them or gathered in their depths into an incredible richness of colour which the calm sea reflected, and against which the white beds of snow that lay upon them shone forth in purest contrast. I stood on the deck, note-book in hand, keenly alert to every novelty.

"Five A.M. (Aug. 8). We are passing William I. Island, and the sea is opening out, dead calm, with huge ice-masses floating upon it. All the men are asleep below. The insignificant Bastion Islands have come into view, and over them I can see, far away, a low white line above the horizon. It is the snowy covering of Barents Land. An ice-blink has come up unpleasantly near the point towards which we are steering. Seals raise their man-like heads and shoulders to gaze at us out of the wonderful smooth, grey sea, whose quality from hour to hour continues with little change. It is now certain that the ice-pack lies ahead, barring our southward way. Only one chance remains—to attempt the passage of Heley Sound."

This strait is marked Helis Sound on the chart, and sometimes Hell Sound—a not inappropriate title, but it was originally named after its probable discoverer, William Heley, skipper to the Muscovy Company, and, according to Purchas, a voluminous poet. Seldom has the Sound been traversed by ships. Lamont, I believe, was the first to navigate it, but only in an open boat. At the best of times the navigation is most dangerous, and generally it is impossible. The ice-pack, filling the south part of Olga Strait, formed a cape, against whose point our course would have carried us. From this cape the edge of the pack trended away SE. and SW. We bent off down the SW. face, running toward Unicorn Bay, with the pack on one side and Cape Weyprecht on the other. For ten miles or so the sea remained open, then drift ice was encountered, which became more and more thick until progress was difficult. A wooden ship might have progressed much farther, nosing her way between the floes and pressing against them, when necessary, to open a passage. Such manœuvres would have sent the *Expres* to the bottom in half a minute.

The difficulty of the passage of Heley Sound lies in this, that the tide ebbs and flows through it, with a speed of from eight to ten knots, so that the water eddies and boils about the rocky islands, and through the narrow places of the gut. For our little boat these would have been difficulties enough, but when the water was covered with floating masses of ice, each hundreds of tons in weight, which the current would have twirled about and furiously banged one against another, the obstacle became insuperable. At slack water we might perhaps have run through, even as it was, but we could not approach the mouth of the straits save after traversing twenty miles of ice-encumbered sea, down a narrow channel between land and ice-pack, whilst at any moment the pack might have set upon the land, and crushed us without warning.

We returned, therefore, to the cape and ran down the east margin of the pack, closely following its edge and searching for a lead. But the ice-sheet was solid, and stretched, practically unbroken, away to the remotest distance. On one side of us was the calm sea with the great bergs floating in it, on the other the white ice-sheet, broad, blue-edged, cracked here and there, and sometimes broken into long high ridges of piled blue and white masses, where two floes had been driven together and "screwed" their tortured edges up into a splintered chaos. Snow lay thickly on the ice-sheet and masked the old ice-heaps; only the ridges recently formed preserved the sharpness and colour of their component fragments. At one place, three walruses lay upon the ice, but they were far off and seemed mere dots of black to the naked eye. We had no time to attempt a stalk, for now, though sea and air were still, it was evident that the wind was coming up from the south, and would probably soon set the pack in motion and plug the entrance to Hinloopen Strait. Farther east we might still find a southward lead, so on we ran, following the pack's edge, which kept trending more and more to the east. Barents Land and Edge Land receded and sank; North-East Land was all but gone—only a white line of gentlest undulation still marked the position of the inland ice. The sky was dark overhead, and the calm water mirrored the gloom, in which the floating ice-castles seemed to hang suspended; but at a short distance they also melted into the general grey.

Now, away to the eastward, a new land rose above the horizon and riveted our gaze. It was Wiche Land, or King Karl Islands. Walrus and whale hunters have known of its existence from early times, but few travellers have ever beheld its rugged crest, most of them only from high points of view on Barents and Edge Islands. In 1872 the crews of three walrus sloops actually landed on one of the islands and brought home a rude description of it, whilst in 1884 Bottolfsen, likewise in a sloop, sailing from the Ryk Yse Islands with a fair wind, came to Wiche Land in three watches, and sailed all round it. Such chances are rare. Eagerly, therefore, I watched the clear-cut white outline of snowy hills, buttressed with rock-walls, coming nearer and growing more clearly defined. Following the pack's edge we were heading almost directly towards it. We might perhaps have actually steamed to it through open sea, but still our main desire was to find a gap in the ice-barrier, and extricate our boat by passing south before the ice closed in again upon Hinloopen Strait.

There was no southward opening. The pack remained continuous, stretching right away to Cape Walter of Wiche Land. Only in the sea the loose floes became more frequent and caused us to make a zigzag track. At last we could advance no farther. The way was barred across and we had to stop. It was still dead calm on the water amongst the floes. Clouds and besoms all around were drifting away before the south wind, leaving us no choice but to return as we came and with all speed, for a few hours of the south wind would suffice to drive the fast ice up against North-East-Land, and cut us off from all chance of escape.

A few minutes we lay on the still water in the midst of the silence and the ice, examining with telescopes and photographing Wiche Land, now so clear and tantalisingly nigh. We could trace the fashion of its architecture, and recognise the bed of hyperite that crowns its northern cape. To the south it seemed to slope down to a low white line resting on the horizon. One more climb to the mast-head in hopeless search for a lead; then round goes the helm, the engines throb, the water boils under our stern, and we are off again for Verlegen Hook, a straight run of about 125 miles. Would the ice there permit us to pass, or should we find ourselves cut off? That was the question which only the event could answer. The cold air, pouring over the stern of the boat and down the companion, banished warmth from the little cabin. By good luck the wind was coming up strongly from the south; a south wind would tend to keep Verlegen Hook clear. That we were retracing our way had this merit, we could sleep and lose nothing. Off Cape Torell the sloop *Gottfried* of Tromsö was hailed. She said that, three days before, the pack had been fast across Hinloopen Strait at this point. It opened out suddenly and was already closing again. Another sloop was seen a few miles away, hunting in the ice close under North-East-Land. Both followed us up the Strait.

A strong current with us and a strong wind behind made our passage quick. At four A.M. (Aug. 9), I awoke to find Treurenburg Bay close at hand, and the sea to Verlegen Hook open, with no ice-blink over it. The water was calm under the lee of the land and everything was grey, but not so dark as before, nor was the cloud-roof so low or so heavy. Otherwise the character of the scenery was the same. There was no sign that even a partial eclipse of the sun was taking place; the normal gloom reigned. Again ice-masses dotted the steely sea, and an ice-blink shone across the north, where marshalled storm-squadrons, rolling along like the smoke of guns, magnified by their complexity the aspect of the cloud-roof, which here was broken by a "lead" of light, there dropped a tender veil of grey upon the level sea, or mimicked the forms of mystic beasts with the flying offspring of an errant fancy.

We were passing Verlegen Hook without mishap, keeping as close to the shore as the shallow water would permit, for now our destination was the secluded Wijde Bay, whose recesses, rarely ice-free, have seldom been penetrated by explorers. The faint low line of North-East-Land was being hidden behind the near and dark low spit of the Hook. The bolder front of North Spitsbergen was ahead, and close at hand were the craggy shapes of the mountains ending in Grey Hook. A cloud-roof rested on them, and they were white to the base with a heavy covering of new-fallen snow. Fog, trailing skirts upon the water, crawled out of Wijde Bay and Hinloopen Strait. A brighter pallor glimmered in the southern sky. Cold steely gleams came and went in this place and that. Seals, rising and diving again, made rings upon the calm water. Hurrying bird-flocks skimmed the surface, guillemots with fish in their mouths for the young ones at home, little auks,

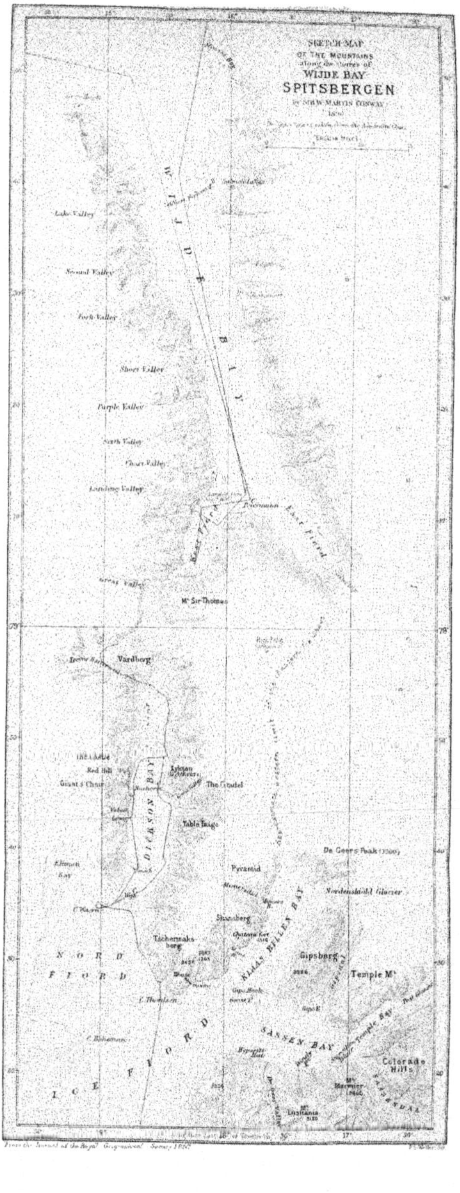

From the Journal of the Royal Geographical Society

GREY HOOK FROM WIJDE BAY.

and all the usual company. Now opened Mossel Bay, where Nordenskjöld wintered; we could distinguish the position of his house. What a cold dead world it was! with the strong white near at hand, the warmer cream-tone farther back, and the bluish-grey beyond—lovely but undesired, like the beauty of the grave!

Everything on this little voyage happened too fast. Events followed one another too quickly. I have to tell in a chapter a tale that should fill a volume. Time enough, indeed, there was in which to gather a definite impression of each scene. Hinloopen Strait, Wiche Land, the Seven Islands, Olga Strait, Wijde Bay—they are all quite clearly photographed in my memory—clearly as Zermatt or any other place in which I have spent months. But between the pictures were no proper gaps, such as in the rest of the world night provides. They were unframed, reeled off on an endless tape. Thus my story suffers. Could we have halted and slept at Mossel Bay, then gone on board again and made a new start, the voyage up and down Wijde Bay would have appeared a thing apart, a separate experience, not a mere incident in the doings of a tenfold day.

The striking peculiarities of Wijde Bay are its straightness, depth, and general uniformity of width. As we entered, the sun for a moment shone, but for a moment only. The Grey Hook hills were clear, and so heavily encumbered with new snow, that it was with difficulty if a rock here and there struggled through in the steepest places. Opposite are curiously-knubbled hills, of hard gneiss, I believe, rigorously glaciated. It was not till Aldert Dirks Bay was passed that the grand scenery of this noble fjord was really displayed. Had the vista been clear to the end, on our entering, as it was on our retreat, we should have received at this point a grand impression of the glories of the fjord.

As it was they only came upon us with slowly-accumulating force, for a cloud bulged down on to the water a few miles on and blotted out the end. All along the west side stand a row of fine mountains, mountains increasing in majesty of form as one advances south. They are penetrated by numerous valleys, some long and level, others short and steep, others again dividing into many branches of attractive topographical complexity. All are lonely; most have glaciers plastered on their sides or at their heads, but the floors, except near Grey Hook, are usually flat and boggy, like the floor of Advent Vale.

The new snow, lying down to sea-level, only lasted a little way into the bay. Farther in the snow-line rapidly rose to about 1000 feet above sea-level. The north coast and its immediate neighbourhood seem to have a much heavier snow-fall than the interior. The aspect of the deep snow-beds covering valleys and plateaus, the snow-slopes and small glaciers, and the general barrenness all along the north, show that the snowy condition is habitual. Equally habitual seems to be the more temperate weather a few miles south, for there is a fair amount of vegetation on which plenty of reindeer support themselves. The contrast is most striking; not a few miles only, but several degrees of latitude might be imagined to separate these neighbouring zones. The coast is the most Arctic region we saw; the bay appears relatively temperate.

The mountains along the west side of the bay are, as I have said, magnificent. They resemble one another in type, and so do the valleys that sunder them. The east shore is less uniform and less generally fine, but the average of its scenery is raised to a high level by certain splendid prospects. In the main, as already stated, the New Friesland peninsula, dividing Hinloopen Strait from Wijde Bay, consists of a high icy plateau with a glaciated rock-front

THE WEST SHORE OF WIJDE BAY.

GLACIER IN THE EAST SIDE OF WIJDE BAY.

on the west. The ice stops short at the edge of this glaciated wall, except where it finds its way down ravines cut down into the rocks. Of such ravines there are three large and as many small. Down the large ones flow wide ice tongues, gently sloping, very smooth, and offering a notable contrast in texture and colour to the richly toned and buttressed precipices of gneiss and ancient schists that wall them in on both sides. Their crescent fronts, expanding beyond the mouths of their ravines, rise like walls of marble and turquoise from the dark clear waters of the fjord. The delicate sky line of the inland ice above completes the picture. Opposite the second glacier we had the steamer stopped in the midst of this scene of rare beauty, just as the sunshine fell upon the splintered glacier-front and the calm water, whilst the red, riven crags remained behind in gloom. An absolute stillness reigned. Seals peeped forth around us, numbers of fulmar petrels floated upon the water near the boat. The clouds, resting on the snow peaks opposite, moved and somewhat opened, showing the direction of white ridges, and the existence of sharp summits, whilst, up a deep and many-branching reindeer valley, which opened just over against us, was such a deepening and enrichment of colour to an incredible purple in the remotest depth, as I think can seldom have been surpassed even in an oriental imagination.

Garwood, whose alertness never failed on land or inland waters, took advantage of the pause to feed and photograph the fulmars. They squealed and quarrelled over the biscuits he threw them, and for a moment, a score of them, forgetting their shyness in the excitement of combat, came close under the steamer's bows. Ted and Battye industriously sketched, complaining of the rapidity wherewith lights and colours changed. The cloud roof grew thinner, and there came a shine and glamour on the snow, and in the wreathed avenues of mist. Bluer grew the ice fronts, ruddier the rocks, more purple the valleys, and more green the sea.

Round went the propeller again, lining the water with diverging waves. The panorama went on unrolling. Devonian rocks invaded the west shore and streaked all the lower slopes with skirts of red débris, which, buried in blue air, produced astonishing effects. The east wall sloped, gentlier and unbroken, up to the ice-field, over which domed clouds rolled along. Farther in, still on the east, came deep cirques, surrounded by fine arêted peaks of hard rock, all white with new snow, and jutting their heads into cloud. And now we were approaching Cape Petermann, the bold, steep-sided, mountain promontory that divides into two fjords the head of the sound. The east fjord is the true continuation of the main valley, and retains its leading characteristics. It terminates in a broad, smooth glacier, which flows down from the inland ice, and appears to be advancing and driving back the sea. Narrower and grander is the west branch, in reality a side valley, depressed below the water level.

We turned up it, minded to land on the promontory in the interests of botanical collecting, but the geologists, preferring the Devonian of the west coast to the Hecla-Hook of the Cape, gained the day, so we crossed over, going very slowly and sounding, for there is little depth.

During the day Trevor-Battye collected several birds, of which he was in need. At this point he shot a young little auk. A greedy glaucous gull instantly snatched it up, but the other barrel dropped him stone-dead, amidst cries of "Sold again!" from all on board. When the others had landed, I went on alone up the bay, but with little profit, for its head is quite shallow, and nothing new came in sight.

WEST FJORD, WIJDE BAY.

MOUNT SIR THOMAS AT THE HEAD OF WEST FJORD, WIJDE BAY.

The land is rapidly encroaching on the water, as at the head of Dickson Bay.

We saw no glacier emptying into the bay at its SW. extremity, though one is indicated on the chart. It may have been hidden behind a corner. Pedersen said there never was one here but a long green valley, far up which he once went in pursuit of reindeer ; we had learned not to put much confidence in his topographical memory. The valley, whether containing a glacier or not, runs in westward behind a fine sharp peak that stands out alone and forms the most striking object in the view. South, at the head of the bay, is a large mountain mass, to which I have ventured to give the name Mount Sir Thomas. Wijde Bay, like the greater part of the coasts and bays of Spitsbergen, was first explored by English whalers. They named it Sir Thomas Smith's Bay.

The Dutch, following the English, called the place Wijde Bay, and that name has stuck. It is a pity that old designations should be thus obliterated. Sir Thomas's name should, at all events, be preserved somewhere in the neighbourhood. Most of the Norwegian, Swedish, and German names that figure on our Admiralty Chart, and receive currency from it, usurp the position of the original English or Dutch names. The substitution for Wiche Land (so named by its discoverer, Thomas Edge) of the modern designation King Carl Land is the most glaring instance of this impiety.

East and west of Mount Sir Thomas are passes by which access might be attained to glaciers leading down to the swamp at the head of Dickson Bay. The east pass was reached from the south by Lieutenant Stjernspetz in 1883. Clouds prevented us from seeing it, and I was not then informed of its existence, or I should certainly have landed and explored it. Mount Sir Thomas might have been ascended from the col, but what was the use when, save for a few moments, it was buried in clouds almost to the base, and so remained as long as we were in Wijde Bay? The water in all this inner part of the fjord, as at the head of Dickson Bay, is of a reddish mud colour, stained by dust of Devonian rock. We lay still awhile at our turning point, and the sun again shone forth, casting an island-mantle of radiance into the midst of the solemn region. Hither it was well to have come. Nature doubtless in many places builds her secret and almost impregnable fortresses on a bigger scale. Many such were known to me in other parts of the world. Memory now recalled them in order—her most precious possessions; but I was obliged to admit that among them was none in which all the parts worked more admirably together to produce an effect of grandeur than in this remote and secluded fjord, with its bare red and purple hills, its snow slopes and bulging glaciers, its roof of uneven grey and white cloud, its calm dark waters, its stony shores, and the broad beam of sunshine striking low.

We returned and gathered up the others, then ran across to Cape Petermann, and, taking our departure thence, made a straight run down the midst of the bay. The engineer undertook to keep his machine going at a uniform velocity; thus, by noting the times and taking a succession of bearings, I was able to make a rough sketch survey of the hills along both sides. A lace-like curtain of mist drooped over Petermann Hill as we ran away from it and into the freezing wind. For five hours I stood at the head of the companion taking observations with fingers deadly cold. The scenery was as lovely as before, but I had eyes only for topographical facts and must be blind to effects of beauty, collecting materials for science instead of art. So absorbing was the occupation that the five hours passed almost like a moment, but when Grey Hook and the point off Mossel Bay were in line with our boat and my work was done, fatigue and hunger made themselves felt, and demanded satisfaction.

WEST SHORE OF WIJDE BAY.

THE THREE CROWNS FROM KINGS BAY.

CHAPTER XXII

THE WESTERN BAYS OF SPITSBERGEN

TURNING sharply to the west, at the mouth of Wijde Bay, we headed for the Norways. I just looked into Liefde Bay, with its numerous islands, and surroundings of black and white hills, glanced at the low land of Welcome Point and Red Beach, then went below and fell immediately asleep. The *Expres* travelled steadily along, passing between the Norways, then south of Cloven Cliff and Vogelsang, and so through Dutch Bay to where Herr Andrée's ship lay at anchor in Danes Gat. I awoke to find the *Expres* anchored alongside of the *Virgo* at four A.M. on August 10, after a voyage that was in some respects as memorable as it was delightful and interesting.

A few hours' respite from travel were both necessary and desirable—necessary, because the *Expres* needed coal and the *Virgo's* people were asleep; desirable, because our powers of observation were jaded with new scenes, and needed time to recruit. At the *Virgo's* breakfast hour the skipper came over to make inquiries, on the part of Herr Andrée, as to the winds we had recently encountered. We could only report a tolerably continuous breeze from the south, except for the local northerly blast up Wijde Bay.

It was certain that the general air movement over Spitsbergen was towards the north; at that moment, a strong south wind was blowing, and the heavens seemed to be on the point of clearing. Hopes arose that the balloon's hour might be at hand; but Herr Andrée wisely decided that it was not worth while risking success, so late in the year, for the sake of a second-rate chance. Even if the wind was good, the weather was unsettled and far from clear. A weaker man would probably have gone up; to my thinking Herr Andrée, of whose determination to ascend when the right time came I was and am fully assured, gave evidence of real strength and judgment, in preferring to return home and face misunderstanding rather than to go up at an unsuitable time of year, and in weather that was certainly doubtful.

Presently Mr. Stadling came on board and gave us all the news. We took on the mail for Europe, and the coals. At ten A.M. the anchor was raised, and we were off. "We came in by the South Gat; which way shall we go out?" was the question. I chose Danes Gat and the rough sea at once, for novelty's sake, and that we might look into Robbe Bay, now deserted as the Seven Islands themselves, but in old days thronged with whale-fishers. The sun at last shone warm again as we entered the heaving sea, the south wind blew so strongly and all looked so fair, that we thought of turning back to tell Herr Andrée how things were, for in the enclosure of his rock-bound bay the state of the weather at large was hardly discoverable. It was decided to mind our own business, and on we went. Robbe Bay, with its ice-smoothed hills of hardest rock, was quickly passed. Then we saw ridges decked with snow and crested with needle rocks shining in sunlight over the end of Danes Island. They belonged to what the old navigators used to call

Knottie Point, but this original name, like most of the other English names, has been wiped off the chart.

Keeping far out, to avoid certain sunken rocks, till we had come opposite South Gat, we ran for Magdalena Bay. Its fame is indeed well deserved. The snow-fairies possess few lovelier retreats in all the glacier regions of the world. There was no time to spare, for the swift south gale was bringing up a storm of snow which came with the enveloping clouds, and was about to overflow the hills into the bay. The heaving sea, the dashing water, the tumbling boat, the noisy wind, the low-racing clouds, tearing down over the hills like some tidal wave breaking on a rock-bound coast, formed a combination of circumstances which gave to the view of Magdalena Bay a setting of excitement. There was a vision of purest glaciers, tumbling between jagged and jutting ridges of aiguilles, seamed by steep couloirs, a vision of glaciers with wide blue-fronted snouts washed by the blue sea, a radiating series of ice-rivers and cataracts stretching back on all sides into the icy interior. The buttress arêtes were like so many Peteret ridges, powdered with fresh snow on such little ledges as would retain it. A moment only we beheld these things; the curtain descended, and they were gone.

Round went the helm once more, and back we plunged into the open sea, now white with foam-caps, which the wind caught up and drove along through the air. The Seven Icebergs and all the features of the coast were lost in fog, and a painful time we had of it for the next few hours till Mitra Cape was passed. Here Kings and Cross Bays open from a common mouth, and we bent in to see them. Cross Bay is a deep and gloomy retreat, now seldom visited, but once haunted by whales and their captors. Hills rise steeply on either side, and seem to multiply in the distance, where the end splits into three small divisions. Hill-

MAGDALENA BAY.

GLACIER AT THE HEAD OF KING'S BAY.

sides and valleys were white with snow. If Cross Bay presents an effect of depth and narrowness, Kings Bay, without really being much wider, has an aspect of breadth. This is due, partly to the fact that the mountains do not rise directly from the water, but from the head of a wide sloping beach; partly because the end, instead of being enclosed by a throng of peaks, is occupied by the broad front of a splendid glacier, no less than six miles wide, flowing from the south-east and draining the same reservoir which supplies the great ice-tongues that empty into the north side of Ice Fjord and round as far as Dickson Bay. From this glacier rise the famous peaks called the Three Crowns. These, according to the map, stand at the very head of the bay, and should have been visible from its mouth. I was puzzled by their non-appearance. In fact, they are not situated in the position assigned to them, but farther north, their direction being indicated by this, that when we were about two miles due south of the Middle Hook dividing Cross and Kings Bays, the three peaks were hidden behind the rounded promontory of red glaciated rock which forms the east side of Blomstrands Harbour. The Crowns did not appear till we had cleared that promontory, and then not three in a row, but at the points of a right-angled triangle.

Again came the luck of a little sunshine, casting upon the icy bay a brief aspect of gaiety. The wide glacier, stretching back into the expanse of the interior, and burying the bases of a number of considerable hills, was the centre of attraction. On the north were splintered peaks of archæan formations. The Crowns and their companions are built of more recent rock, horizontally bedded, and which, by its aspect of regularity, renders the hills like the work of men's hands, so that their relatively small scale as natural objects is forgotten. Each of the Crowns resembles the second Pyramid of Gizeh—the one still capped with the remains of its original casing. Like it they are ruinous, but strangely symmetrical, peculiar rather than beautiful. In this strange region of unusual forms they are admirably placed, with splintered peaks near at hand for contrast, the huge ice-rounded mass of hard red granite close by, and the ice-sheet like a great marble floor below, between, and beyond them.

In the great ranges each component mountain is so large that a few only can be seen at one time from near at hand. From high and wide-embracing points of view alone can the effect of complexity be obtained. But here, where each mountain is actually small, though of noble proportions, many can be seen at once from no great distance; thus each view has an involved appearance, for there are always several whole ranges to look at, and not merely a few mountain units. So it is at the head of Kings Bay. One looks up several valleys, some of them so deeply filled with ice that arms flow across from one to another, actually isolating the peaks and turning them into true nunataks, the only true nunataks we saw from near at hand. If the glaciers were to retreat, the bay would extend much farther inland, for the whole of its head is filled up with ice, to what distance there is no means of judging.

Singularly inviting as a highway was the chief glacier, stretching back SE. to an apparent col, whence the great glacier of the North Fjord probably takes its origin. In fact, we saw no place better suited to be a centre for the exploration of the ice-sheet and the mountain ranges than Kings Bay. A mountaineer, with the existing facilities, might be landed here with a sledge and the requisite appliances, and left for a month. In that time he could do most valuable work, besides having delightful and novel experiences. Let the Alpine Club look to it; the novelty will not last long.

After drinking our fill of the fine prospect, we turned about and headed for the sea, making our way through many great masses of floating ice—seracs fallen from the glacier cliff, and already melted into the most fantastic forms. One was of such beautiful outline and colour that we came to alongside of it to give the artists time for sketching. Then away to the sea, and the recongregating fogs and clouds, which utterly hid the end of King Charles's Foreland. Once round Quade Hook wind and waves were dead against us. The mountains on either hand of Keerwyk were invisible. Only a narrow strip of beach or rock divided the watery heavens from the misty sea. Landmarks were hidden; the boat had to be steered by dead reckoning through the shallows. As there was nothing to be seen we turned in, and only the skipper kept the deck, planted in his little elevated steering-box just forward of the funnel. A rude shock fetched us suddenly from the land of dreams. We had touched bottom on the Bar, the bank which Barendsz discovered and could not sail over. A second and more prolonged bump followed, then a third. Our strong little craft gave a kind of jump, and we were over, lurching and rolling as before. A few moments later all were asleep again.

The orders were to run straight for Bell Sound, passing the mouth of Ice Fjord. Gregory was to be landed at Cape Lyell, where is a famous bed of fossil plants. Garwood and Trevor-Battye desired to explore Axel Island. I wished to see the head of Low Sound, and to have another look up the valley of the Shallow River, visited from Cairn Camp by Garwood and me on June 28. It was about seven A.M. (Aug. 11), when we awoke off the mouth of Bell Sound and prepared to put our plans into execution. Clouds were on the hill-tops, but below them the air was clear, and the eye ranged afar. I had to rouse myself to observe, for I was becoming tired of the sulking cloud-roof, the blustery chilly south-east wind, and the splashing sea. Restless water in front, a band of black and white beyond, and then the long grey firmament—the combination was becoming wearisome. I longed to be on solid earth again.

When Gregory had landed, running was made for the north end of Axel Island. This long, low, carboniferous rock, rising at its highest but a few metres above sea level, and separated at both extremities from the land by narrow channels, almost blocks the entrance to Low Sound. At one time it must have been continuous with the hills north and south. The ancient glacier that filled Low Sound perhaps helped to tear it down, taking advantage of the vertical dip of its slab-like strata, so similar in arrangement to the rocks of the Karakorams, which ice also breaks away when it flows across their strike. When the Low Sound glacier departed, water denuded the soft horizontally stratified rocks and cut gorges at both ends of what became an island when the whole was depressed, and the sea ran in and filled the valley.

It is for geologists to say whether Axel Island corresponds in nature, as it certainly corresponds in form, with the transverse submarine rock-bars across the mouths of several Norwegian and Scotch fjords. Once this island must have been below sea level, and formed just such a submerged bar. It has risen above the surface with the general elevation of the land now observable in Spitsbergen. There is a similar though less level-topped bar across the mouth of Van Keulen Bay. If the early charts are to be trusted, the channels north and south of Axel Island were considerably wider in the seventeenth century than they are to-day.

All the bays in the west coast of Spitsbergen are submerged river valleys, and the rivers that formed them had, as they still have, easy work in cutting down the soft horizontally bedded rocks that form the bulk of the interior. But along the west shore they were confronted by the line of hardest and most ancient rocks. Relatively narrow exits therefore are the rule, exits sometimes used in common by two or three main valleys. At either side of these the ends of the hard ridge stand up as monumental doorposts of the gate. Ice Fjord owes much of its fine first impression to the peaks of the Dead Man and Mount Starashchin, between which it is entered. There are corresponding Pillars of Hercules to Low Sound. That to the north was most striking. I have been told that its proper name is Bell Mountain, and that from it the bay is called Bell Sound. I have seldom seen a mountain that, for its size, impressed me more. It possesses, from some points of view, an aspect of bold and curving uplift, emphasised by its shaggy head and great mane of snow. The sweeping outlines of its clearly defined and contorted strata, with the strange and varied colouring of the rocks, and the brilliant green on the ledges, all the lines leading up to the bold, nodding brow, form the elements of a mountain composition of excellent quality.

The tide was setting in through the narrow entrance with a great rush and turmoil. A sound of many waters filled the air, a deep diapason extraordinarily impressive. The otherwise smooth strait bulged up from below in swelling domes, round which wavelets broke in rings towards the centre. Rushing currents caught and carried the steamer this way and that, and swung her about like a cork. It was all most agreeably exciting.[1] Once within the gate and under shelter of the island all was calm. Garwood, Trevor-Battye, and Ted went off in the row-boat to land, whilst I steamed on up the north side of Low Sound. Into the south side thirteen parallel valleys empty. They are all much

[1] It is stated that the tide runs much more strongly out than in, as might be anticipated, and is more rapid through the south than the north entry.

alike, and are divided from one another by regular, flat-topped, even-sided ridges, like so many railway embankments, end-on to the fjord.

The north shore is more varied and beautiful. It begins with a wide green valley, debouching west of Coal Mount, and apparently continuing the depression of which Green Harbour is the north extremity. Looking back now upon Bell Mountain, whose form as beheld from this point of view matches its designation, I thought I had never seen a more splendid scheme of purple colouring—grey purple were water and sky; the glacier front and slope were of a darker tone, whilst the rocks and low land, Axel Island and all the shore, were so rich as to be almost like a warm black velvet. In the clouds over the snow was an ice-blink, adding an element of weirdness, but the grandeur of the mountain dominated the whole.

It is needless to catalogue the glaciers and valleys that diversify the north coast. They were of minor interest to me. One of them continues the Coles Bay depression, and all give access to the green and boggy interior, which is only snow-capped at a relatively high level. What I was on the look-out for was what the unimaginative Norwegian hunters call the Stordal, which empties from the north-east into the northernmost bay of the fjord. It was named Ondiepe or Drooge (Shallow or Dry) Rivier by the early Dutch navigators, and the name ought to be preserved. Dreary Valley empties into it. It was the mouth of Dreary Valley I was now anxious to see, in order to verify a bearing taken thence on that dreadfully cold June morning, and which I had since imagined (wrongly, as it proved) to have been inaccurate. Rounding a low promontory, the *Expres* came in sight of the wished-for view, and at that moment her engines broke down. I was not sorry. Notable scenes had been reeled off before our eyes so rapidly, and for so many successive days, that it was pleasant to have pause and lie still in one place, if only for an hour or two.

I lay about amongst the coal-tubs on deck, and drank in the view at leisure. The chilly air presently drove me below into the deserted cabin. I measured and made a plan of it. In area it was about twice the size of a grand

BELL MOUNTAIN FROM LOW SOUND.

piano where it was broadest, that is to say, above the shelf! Floor, seats, and shelves were occupied by a chaos of things—bottles, sleeping bags, cooking utensils, spirit tins, baggage, coats, rugs, botany presses, various tins, cameras, charts, sketching blocks, changing bags, geological hammers and specimens, instruments, guns, cartridge-boxes, books, food, plates, and countless boots, long and short. This heap was about knee-deep, and lay just where the motion of the boat had rolled its component parts. The confusion was not unpicturesque. I did not at this time learn what we were afterwards told, that the engines had been all along in an extremely "dicky" state. The boiler had not been cleaned for a long time, and was thickly encrusted. Various bearings were coming loose, and there were many other matters requiring attention. The fact that we were not paralysed in the midst of the ice is not the smallest piece of good luck that attended our exceptionally lucky voyage.

In due time the machine began working again, and the boat advanced over the now glassy fjord. Quite small waves, rendering uneven the surface of the sea, are large enough to hide from view birds floating or flying low beyond a very short radius from the deck of a little boat. But when the water is calm, the range of vision for objects upon it widens, and a small disturbance of its smoothness, even at a great distance, immediately attracts attention. Now the whole visible area seemed alive with sea-fowl, floating, rising, or diving. Fulmars were specially in evidence, and I noticed that, when rising off the water, they run rapidly along the surface, striking it hard with alternate feet in a manner adopted by no other bird in these regions.

After approaching as close to the mouth of the Shallow River as we could without going aground, and when all the information I required was obtained, we bent away southward under the end of Sundewall Mount, and turned up the extreme easterly branch of the fjord. The scenery became grander, a delightful harmony in blue with just one white curtain of mist, let down over the hill crests and the deep indigo lower slopes from the blue clouds to the blue water, whose still surface was broken by the slow emergence of many seals. The skipper shot one, so fat that it floated and was hauled on board. The little deck ran crimson with its blood.

It was annoying that all the larger mountains should have remained invisible. Fox Peak and its neighbours, so well known to me, should have been in sight, but they

VALLEY OF THE SHALLOW RIVER, FROM LOW SOUND.

were cloud-capped and fog-enveloped, and so remained. After completing the tour of the bay and taking the Axel Island contingent on board, we ran for Gregory. The water being slack, our exit through the narrows was unexciting. Gregory came off with a boatful of fossil plants. He had beguiled some of the waiting time by a long talk with Svensen, a most intelligent sloop-captain, whose vessel was at the moment in Schoonhoven, with an almost full cargo of white whales. The wind had dropped, but a most uncomfortable ground-swell made it difficult to lie on the narrow seats. The *Expres* rolled intemperately about. All of us were afflicted with violent colds—an epidemic probably caught from the people of the *Virgo*, for we had been free from such trouble while we remained alone, notwithstanding our constant wettings and chills. The balloon colony were in frequent communication with Europe, and were sufficiently numerous, perhaps, to keep colds going among them. Whatever the cause, sneezing and coughing now raised echoes in our little cabin, and added to the discomfort of our passage.

Just as we were quitting Schoonhoven (Aug. 12), Garwood suggested that we should run south, and visit Horn Sound, for the purpose of climbing the highest measured peak in Spitsbergen. It was originally named Hedgehog Mountain by the English whalers, but the Norwegians have changed this to the commonplace Hornsunds Tind, by which designation it is now generally known. The ascent was part of my original plan, and I would gladly have gone to make it, but duty called me to Advent Point, whose latitude I still needed to observe, as well as the true bearing from it of Bunting Bluff. The date fixed for the departure of the last steamer was August 15, and all my efforts to make arrangements to lengthen our stay proved unsuccessful. Northward, therefore, the *Expres* had to be steered.

Printed by Lemercier, Paris

ADVENT HILLS FROM ADVENT POINT

CHAPTER XXIII

HORN SOUND AND HOME

EARLY on August 12, Starashchin Cape was rounded and Ice Fjord entered in a dense fog, through which the way had to be felt to Advent Bay. At seven A.M., the anchor was dropped in the old anchorage, and we landed, after a most interesting voyage of over 1000 miles—the longest in so small a steamer ever performed in Arctic seas, and I believe, after searching all available records, the most complete coasting voyage around Spitsbergen ever accomplished in a single summer. Goods and kit were landed and camp pitched by noon, with the accuracy that comes from practice. The tents were within a few inches of high water mark, and the day's spring-tide almost flowed in. A pallid flush of sunshine gleamed for a moment, and Bunting Bluff almost became visible, but the clouds descended before the sun could be seen, and a steady downpour of rain, thoroughly English in type, set in. Garwood now came to me to say that he had arranged with the *Expres* to take him to Horn Sound, to climb Mount Hedgehog. The engines needed some botching, but the boat would be ready to sail next day (13th), and they would reach Horn Sound on the 14th. The *Lofoten*, the steamer that was to carry us to Norway, was to sail on the 15th. With luck, therefore, they might climb the mountain on the 14th and 15th, in time to run out with the *Expres*, and tranship to the *Lofoten* in the open sea. Failing that, the voyage to Norway must be completed in the *Expres*, which of course had to return to Tromsö somehow. Trevor-Battye, Garwood said, had agreed to come with him, though he was not eager to climb the mountain, having never made a regular mountain ascent before.

Till the sun had been observed it was not possible for me to go. My presence on the mountain was in no sense necessary, as Horn Sound and the surrounding hills had already been mapped by Von Sterneck in 1872.[1] Moreover, there was the business of the expedition to be attended to, and the remaining baggage to be packed. I was pledged to return with the *Lofoten* to look after the sale of the ponies at Tromsö and other matters. I was, therefore, forced to let Garwood and Trevor-Battye go without me. Garwood heard my decision with heartless glee. "Then you will lend me your climbing boots. I sent mine back to Norway by the last steamer along with the other baggage." It really showed wonderful self-denial on his part not to have told me this sooner, for without nailed boots he could not safely have made the ascent, and mine were the only ones to be had.

We spent the rest of the day putting things together for the Horn Sound party. In the evening I had a long talk with Bottolfsen about Spitsbergen traditions, the Russian trappers, and the doings of Mr. Jeaffreson's party while he was with them, all which I duly noted down from his lips. He said that the last Russian hunters who visited Spitsbergen were a group of three or five, whose huts were on the Dun Islands. They met a tragic end, which shall be recounted as far as possible in his own words.

"The story," he said, "is written down and printed. It is well known in Hammerfest and Tromsö. I once read it and have often heard it told, but I do not now remember all the details. It was, at all events, to this effect. There was at Hammerfest a skipper named Andersen, by birth a Dane, but regularly settled in Hammerfest. This year—it may have

[1] "Petermann's Mittheilungen," 1874, Tafel 4.

been fifty years ago, or more—he sailed with his sloop in the spring, and came in June to the Dun Islands. Now the Russians had been very successful in their winter trappings, and they had a great quantity of skins, which Andersen saw and coveted. He thought it would be cheaper to take them than to buy them, so he just killed the Russians, who were weak, and took their stuff away. He killed them with a harpoon on which was his name, and, when he went off, he forgot the harpoon and left it behind. Shortly afterwards the skipper Stuer of Tromsö came that way with his sloop, and he too landed on the Dun Islands, and found the bodies of the murdered Russians, and in one of them Andersen's harpoon sticking, so he knew what had happened. He sailed away and met Andersen's sloop, and went on board and talked with Andersen, who suspected that Stuer had found him out, though nothing was said. At all events, Andersen was afraid, and considered how he might be rid of Stuer.

"They sailed on, hunting along the edge of the ice-pack, and one day, when they were very far from land and Stuer was away from his sloop in his walrus-boat, Andersen went on to Stuer's sloop and managed to do it some harm, so that presently it seemed to be sinking. Then he went again to the sloop and rescued Stuer's wife and the people on board, and sailed away with them to Hammerfest; for, what with the things he had taken from the Dun Islands, and the catch he had made, he had already a full cargo. At Hammerfest he landed the people and his cargo, and told how Stuer's sloop had gone down, and how Stuer himself must be lost, for he was away in his open walrus boat, and had not returned and could not be found. Then he sailed away again from Hammerfest to the ice.

"Meanwhile Stuer had returned to his sloop and found her in a bad way, but he succeeded in patching her up and brought her back to Tromsö, where he met his wife. He soon saw what Andersen must have done, so he related all that he knew about the Russians. But vengeance was already on Andersen's track. He took his sloop far up into the ice, which came packing all around him so that he could find no way out. Leaving the ship, he got on to a high iceberg and climbed to the very topmost peak of it, for it was tall and sharp. As he stood on the top, looking all round for a way to come out of the ice, the great iceberg trembled, and then turned right over. It flung the murderer into the sea and sucked him under, so that he was never seen again, and went straight to hell."

The wind sobbed and howled, as I sat in the tent listening to this tale. The water of the bay broke on the shingly beach a yard or two off. Rain deluged the tent roof. The surroundings matched the story which Bottolfsen told with energetic gestures and glittering eyes.

About noon next day (13th) the *Expres* sailed in the continuing downpour. The outlook was not promising for a climb, but I felt confident that if the thing could be done, Garwood would do it. All day long I worked at journals, records, and baggage. In the evening, now noticeably duller than the day, Gregory and I wandered up to the winterer's grave and the ruins of his hut. Clouds were low over the bay, but a pale misty light crept beneath them, and spread upon the calm bosom of the waters, against which the black crosses stood out sharply. The air was still. A settled melancholy pervaded the silent scene. Darkness hung on the skirts of the hills. A mild drizzle began to fall, and the atmosphere grew thick and greyer. A school of white whales came curving and blowing through the glassy water, close along the shore, where the wrecked cutter lay stranded. Silence laid its hand upon us also. Time seemed to be standing still. It was our last evening on the shores of Spitsbergen.

Next day the *Lofoten* duly came in. She loaded up our things, missing only one case, which, as it happened, was fortunately left behind. In Mr. Jeaffreson's party was, as I have stated before, an artist, Mr. Huyshe Walkey, with whom at this time we had much conversation. Mr. Jeaffreson, who arrived in Spitsbergen on July 10, with the intention of making some explorations, was unfortunately recalled by private affairs on the 25th of the same month, before he had been able to accomplish any new work. He left Mr. Walkey behind, expecting to be detained in Norway for a few days only, and to return and join him. He had not, as yet, been able to do so, but Walkey was under the impression that he would still come and relieve him. Walkey was much urged to avail himself of the *Lofoten's* last trip, and make good his retreat, but he held to his opinion that duty called him to stay, and he stayed. He was insufficiently provided with food, and, had it not been for the case of provisions we accidentally left behind, and which he fortunately discovered, he would have fared even worse than he did. Eventually he was brought home by Baron de Geer, in his already overcrowded cutter.

In the evening came gleams of light in the north, and a stirring in the air. "The weather is going to change," said Pedersen, "See! there is pink light." And now a series of the rarest and most delicately beautiful effects succeeded one another over Ice Fjord, or were simultaneously exhibited side by side to the despair of our artist. The cloud-roof rose just clear of the glaciers of Cape Boheman and their dividing hills, but from it depended a veil of mist, lacelike and transparent, through which the hills were clearly visible. The midnight sun was low, hidden behind the clouds, but its light flowed beneath them, and tinted the mist-veil, hills, and glacier-flat with tenderest rose, except where blue shadows intervened, or purple rocks obtruded.

Each long glacier, sweeping gently down to the fjord, was the centre of a separate picture, perfectly composed. The Nord Fjord's portal was likewise hung with a purple transparent veil, through which appeared ridges of increasing evanescence and incredible wealth of colour, one behind another, till the last melted into light. It was the first sunset glory Spitsbergen had shown us, heralding the long winter night. There was no hurrying excitement of changing effects. Colours and tones lasted long and altered slowly. Over our very heads at length, and away to the south, the clouds became tinged with faintest pink. "Perhaps we shall sail in clear weather," we said, "and see the hill-tops as we go." But it was not to be.

After midnight, when all my work was done and the baggage stowed, I walked over Advent Flats, taking a final survey of the dreary shore. Its flower carpet was faded. All the little plants were in autumnal attire, making sheets of scarlet and orange, gold and brown, on the levels and slopes of the hills. The approach of winter was in the air and upon the earth. Many birds were already gone ; others were going or collecting to go. Our flight-time also was come. The brief summer of Spitsbergen had not seemed brief to us, nor its long day uneventful. Farewell to you, cold and barren slopes, icy broads, bulging glaciers, squdgy bogs, and landlocked waters of so many moods! It is good to have seen you, but a brief acquaintance suffices. Memory will not let your beauties slip. Your toils and discomforts, moved into the past, have become agreeable reminiscences. What we could compass of you we have made our own. Let others who follow learn from our mistakes and profit by our blunders.

After a mid-day meal (Aug. 15), we bade Baron de Geer, Lieutenant Knorring, and Walkey farewell. The anchor was raised and the ship moved slowly along. De Geer's men lit a great bonfire on shore in our honour, bombs were exploded, guns fired, flags waved from the tents as we passed. Walkey gave us a send-off of his own with rifle shots and the dipping of his little flag. A strange passenger on board, in his enthusiasm of recognition, carelessly took a shot at Walkey; the bullet went unpleasantly near. Bullets, in fact, were flying all over the place with really frightening casualness after the true Advent Bay manner! It was all soon over, and we steamed down the fjord in wintry weather with clouds upon all the hills.

On board was a German gentleman, Dr. Wegener, who had come up for a second time this year, hoping to go on with the *Expres* to see Herr Andrée. As the little steamer was absent we ran straight out to sea, where on the far horizon a sail was descried. In an hour or two we came up with the craft, which proved to be a shark-fisher. They catch the fish, the "blind shark" (*Squalus Grœnlandicus*), with a long line, haul them on board, cut out their livers (from which cod-liver oil is made), inflate their stomachs, and chuck them overboard. If something were not done to keep the dead fish afloat, they would sink and no more would be caught. We stopped alongside in the tumbling sea, and the fisher of sharks came off in a boat, and was asked whether he would take Dr. Wegener up to Andrée's. With the usual high estimate of the value of their services which prevails amongst North Norwegian seamen, he demanded 600 crowns for the job, at the outside a two days' sail under the conditions existing. The offer was declined.

We ran down the coast, a long way from shore, seeing only that Bell Sound was filled with a vague leaden cloud resting on the water.

> "The sea-fog, like a ghost,
> Haunted the dreary coast,"

standing out before the island, a leaden wall, through

which one bold ram-like promontory jutted forth. It seemed as though solid night had fallen upon the island, and held it down in a foul embrace. The gloom reached out over the sea, threatening to overwhelm whoever should dare approach. Who could help thinking of "the land of the Cimmerians, shrouded in mist and cloud, where the shining sun never looks down with his rays, neither when he climbs up the starry heaven nor when he returns earthward from the firmament, but deadly night is outspread over miserable mortals"?[1]

By six o'clock next morning (Aug. 16), we were entering Horn Sound, not for the purpose of retrieving Garwood, but to carry off the *Expres* for Dr. Wegener. Snow was falling, and there was much new snow on all sides, yet we had fine views of the crags and glaciers that immediately surround the bay, though Hedgehog Mountain was utterly hidden. In this, as in all west coast fjords, splendid mountains stand on either side of the entrance, the finer being Rotjes Mountain on the north, into the heart of which runs a short green valley. Farther in on this side are two splendid glacier tongues flowing from the inland ice. The right bank of the inner (Paierl Glacier) consists of an exceptionally grand rock ridge, steep on the west, and plumb-vertical on the east side, so that the end section is very bold. At the head of the sound is a third great glacier, and round to the south a fourth, which does not quite reach the water's edge.

Off it the *Expres* was anchored in the retired Goes Haven, and it was up this glacier that our friends had gone to look for Hornsunds Tind, whereof not a trace was discoverable by us. All we could see was the wide gently domed end of the glacier, spreading back into fog with mountain bases on either hand. The rest was left to the

[1] "Odyssey," xi. (*vide* Butcher and Lang).

imagination to picture as it pleased, and in these latitudes it works wildly. The men on the *Expres* stated that Garwood, Trevor-Battye, and Bottolfsen had gone off the previous afternoon, carrying tent and provisions, and had not since returned. In such weather it was easy to believe all manner of misfortunes. Possibly the next few hours would bring them back either successful or defeated. We decided to wait for them as long as the *Lofoten's* margin of time permitted. The anchor was dropped, silence reigned, and snow began to fall heavily.

Gregory, always on the alert, caused a boat to be lowered and went off to dredge. The hours passed and no one came. At noon it was necessary to sail. When we quitted the Sound the surrounding hills were all hidden, and only the base of Rotjes Mount and the broad low front of Torell's glacier could be seen, with the fateful Dun Islands lying before it. In the south the clouds were a little barred and broken with golden lines and the faint reminiscence of sunshine. As we passed South Cape the mists lifted and broke a little, showing the front of the southernmost glacier. Then the peak of Hornsunds Tind peeped through a cloud-hole to see whether we were really gone, and an hour later the whole mountain cleared and stood forth defiant in the north, a mighty tower to all appearance, though really it is but a long narrow ridge, a little peaked at one end. The sun presently reached and began to warm us a little; we faced it with joy, and ran for it, fast as the engines could carry us. Spitsbergen was behind on the horizon of memory. Europe was ahead, and all our thoughts were turned to warmth and home.

During almost two days we were horribly rolled about by half a gale from the west. Then we reached the whaling station of Rolfsö near the North Cape, and Hammerfest three hours later. Signs of rejoicing in the bunting on the ships, and the people crowding expectantly about, were the first information we had of great good news. Nansen had arrived at Vardo in Mr. Harmsworth's ship, the *Windward*, and was expected to reach Hammerfest very soon. The *Windward* had already come in and was anchored not far away. We cast anchor alongside of Sir George Baden-Powell's yacht *Otaria*, newly come in with a successful eclipse party from Novaia Zemlia. Presently the masts of a steamer appeared over a low promontory. It was the mail-boat with Nansen on board.

The population of Hammerfest gathered about the landing-stage. A band played patriotic airs, guns were fired, and the steamer came in, amidst cheers and such enthusiasm as a reserved Norwegian multitude considers it not undignified to manifest. The loudest shouts came from the *Otaria*, and from my excellent German friend Dr. Wegener. A few moments later I was enjoying the great happiness of welcoming Dr. Nansen in person, and hearing from his own lips, and those of his brave companion, Herr Johansen, a brief account, most modestly given, of the wonderful and epoch-making achievements they had accomplished. No more joyous conclusion to our own enterprise could have been desired than this encounter with the greatest modern hero of Arctic enterprise.

CHAPTER XXIV

THE ASCENT OF MOUNT HEDGEHOG, OR HORNSUNDS TIND

BY E. J. GARWOOD

FINDING, on our arrival at Advent Bay, after our voyage round the coast, that three days must elapse before our ship returned to Norway, and that my services were no longer required to drive Carl and the ponies, all of whom had been shipped off to Norway, I proceeded to put into effect a project which I had long cherished, of running down to the south of the island and exploring Horn Sound, and, if possible, ascending Mount Hedgehog. This at least was the excuse I gave for my departure. The real reason was that I preferred being absent when Conway discovered the bill I had run up for luncheons at the Tourist Hut, during my three lonely days at Advent Bay, when I returned to the coast after living for five weeks on Emergency Food in the interior.

Trevor-Battye agreed to accompany me, and Conway lent me his boots. We piled the deck of the little *Expres* with as much coal as she could carry in sacks and barrels, and steamed away, in a drizzling fog, about noon of the 13th, taking with us Ice-master Bottolfsen and a crew of three men.

After an uncomfortable run of eighteen hours down the coast, we arrived at Horn Sound in a thick mist, and anchored among some rocks out of reach of a stream of drift ice. We spent a day waiting for the weather, and were prepared to sit it out; but the weather met us with an equal determination. Our provisions were limited, so we decided to start next day in spite of fog.

The morning was spent in preparations. None of us had much to boast about in the matter of boots. I was wearing Conway's, which were a quarter of an inch too short for me; Trevor-Battye had some old shooting-boots with small mud nails; we rigged up Bottolfsen in some old sea-boots, into which we hammered some cricket spikes, found in Trevor-Battye's canvas waders. There was still another man to provide for, but we only intended to take him as far as our sleeping-place, so I told him to carry an extra pair of socks to pull over his boots. He did so, but as he put them on immediately we landed, he arrived at the edge of the ice with a pair of frilled woollen spats[1] and became rather an anxiety on the glacier. Bottolfsen, having visions of sport, insisted on carrying a gun in spite of our attempt to dissuade him. Leaving the launch at three P.M. we landed in the south-east corner of Goes Bay, which lies about half-way along the southern margin of Horn Sound. Shouldering a tent, a rug, and provisions for twenty-four hours, we started in, nearly due south, over the flat-raised beach, which gradually merged inland into the terminal moraine of a large glacier.

The previous afternoon I had ascended a small hill near the coast, in order if possible to obtain a view of our peak, but everything over five or six hundred feet high was bound in fog, and I could only see the snout of a glacier, which appeared to fill the upper part of a valley. This glacier led in the direction in which, judging by the chart, Mount Hedgehog should be.

[1] In Mr. Garwood's absence abroad I am seeing this through the press for him. His handwriting is not of the most legible. The best the typewriter has been able to make of this sentence is, "he aimed at the edge of the ice with a pair of pilled wooled spots." The above is a merely conjectural emendation !

Lacking more definite information, we decided to ascend this valley, steering generally towards the south-east, and hoping to stumble upon our peak somewhere up in the fog.

After examining the relics of an old blubber-boiling establishment on the coast, we started up the right side of the valley, going sometimes over débris slopes and sometimes on an old lateral moraine; thus we reached the edge of the glacier. On the way we crossed a small medial moraine, composed entirely of handsome grey marble veined with pink, contributed by a small glacier issuing from a steep-sided gorge to the SE. Here we climbed on to the glacier, and entering at the same time into the zone of fog, steered by compass in a south-easterly direction. It was not long before we became involved in a labyrinth of crevasses. They ran, roughly speaking, at right angles to our line of route, so that, prevented as we were by the fog from seeing beyond a short distance ahead, our progress became slower and more laborious.

My companions had scarcely any previous knowledge of glacier work, and it was with increasing anxiety that I watched our friend the cook with the woollen spats cross each succeeding crevasse. After some time spent in endeavouring to reach the left side of the glacier, we retraced our steps to the moraine we had abandoned, and continued for some distance along it till forced again on to the ice. Here, however, the gradient was less steep, and the crevasses fewer and narrower. But the surface was snow-covered, and the bridges treacherous in the extreme. In Spitsbergen, as long as the sun is above the horizon in the summer months, the temperature rarely sinks below the freezing-point in the valleys, so that the snow, left unmelted from the previous winter, is usually rotten and unsafe. Thick bridges, which in the Alps would bear a considerable weight even at midday, here yield readily to a prod of the axe. Extraordinary care was thus rendered necessary; we found it unsafe to jump even narrow crevasses on account of the untrustworthy nature of the landing afforded by their edges.

As we advanced diagonally up the glacier the fog thickened; presently snow began to fall. Matters looked rather hopeless. We sat down on the snow, ate our Emergency Food, and discussed the situation. After a short halt we pushed on again, determined to continue the advance in a south-easterly direction until we reached the watershed of the Island, or were stopped by impassable crevasses. Imperceptibly the fog thickened until one could scarcely see his neighbour on the rope. Although the ground at our feet was so indistinct, that the white lines, indicating the snow-filled cracks of the névé crevasses, were not always discernible, the glare reflected from fog and snow was almost blinding. It was impossible to see at all with glasses, and the strain on the eyes without them was most trying. I felt a curious mesmeric drowsiness stealing over me, such as I have occasionally experienced in a slight degree in Switzerland after a long day on the snow. We were affected by this feeling when crossing the Ivory Glacier, but now the strain on the eyes was greater and the feeling more intense.

Presently we found ourselves again amongst open crevasses which we endeavoured to outflank, but after following their edges for some time Trevor-Battye, who was behind, keeping me straight as far as possible by compass, announced that we were heading due north, and would soon be returning on our tracks. We retraced our steps, turned the crevasses on the far side, and regained the more level plateau. It was subsequently proved that the crevasses ran in a circular manner round the bulging dome of ice, due doubtless to a projection in the floor of the valley at this point. But for the compass we might have wandered round the dome in the thick fog till it cleared the next evening.

We continued cautiously up the plateau after this, prodding at every step, until we estimated that we must have come a distance of at least five miles from the coast. The ground appeared to drop slightly in front of us. We halted to consult, and had just agreed that we must have passed the watershed and be descending towards the east coast, when the fog cleared for a moment in front of us, and revealed within a stone's throw a precipitous wall of rock fully 3000 feet in height. The top was still hidden in the fog but what we saw convinced us that this was certainly part of the mountain of which we were in search. At the same time we discovered a rocky island on our right rising out of the glacier; we decided to camp on it for the night. A bearing of it was quickly taken, and the fog closed down again as thick as before. We reached the rock, however, at 8.30 P.M. without further accident than the temporary loss of a leg in a crevasse. Raking together such stray stones and earth as could be found, we pitched our camp (2074 feet) on a rocky ledge and searched for water. Not a drop could we discover. The quest had to be given up in despair, and our cooking-pot filled with snow. Three mortal hours and a half did it take our little spirit stove to melt and boil sufficient water to mix with our ration cartridges.

At midnight we lay down in a heap on the ground and tried to shiver ourselves to sleep. We had only been able to carry up one thin ship's blanket between us, and the cold was considerable. During the night the wind rose, and snow fell for some hours; the cold made sleep impossible. After fidgeting incessantly for a couple of hours, the cook, who was sleeping across our feet, got up and declared his intention of returning forthwith to the ship. As he would certainly have tumbled into a crevasse before he had gone one hundred yards, we used our eloquence to dissuade him, and eventually compromised matters by allowing him to wedge himself in beside us, though we were already tightly packed. It now became difficult even to breathe; turning round was utterly out of the question. At 6.30 A.M. I could stand the position no longer. I sounded the reveille by a sneeze which effectually aroused my companions.

The weather was precisely the same as the evening before, thick fog and a slight snowfall. Occasionally we could see an outline of the foot of the mountain, and our tracks over the névé nearly obliterated by fresh snow. The thermometer registered nine degrees of frost, but it must have been considerably colder than this during the night. At 9.30 it was still snowing hard, but about one o'clock the fog cleared slightly, and Bottolfsen pluckily volunteered to go back to the ship for provisions and spirit; to my surprise the cook willingly agreed to accompany him. We watched them disappear down our tracks until the fog hid them from sight. Then with a sigh of content Trevor-Battye and I wrapped ourselves in the blanket and dozed till four P.M. We were roused by an unusual glow in the tent, and, looking out, found the sun shining and most of the mountains free from fog. Scrambling into our boots, we ascended to the highest point of the ridge on which our camp stood.

The view was very fine. Bands of cloud were rolling away over the sky, and the sun shone through them. We could see the whole of yesterday's route, and away down in the bay a little black speck indicated the position of the *Expres*. In front of us rose the black and precipitous face of Mount Hedgehog, forming a wall nearly two miles in length, crested by an almost horizontal arête. Here and there a few small gensdarmes protruded. The summit of the mountain rises from the southern end of this arête, where it joins the western buttress, which ran steeply down towards our camp, but, before reaching the névé, this buttress rises again into an irregular ridge connected with

MOUNT HEDGEHOG OR HORNSUNDS TIND.

the rib on which we were encamped. This ridge forms the watershed of the district, and, running round to the NW., ends in the little peak on the shore of the bay, which I had ascended during our stay in Horn Sound. This basin is occupied by the glacier which we ascended from the bay. To the south of this ridge the ground falls gradually to a low swampy tract of raised beach connected with an indentation in the coast-line to the south. It is badly delineated on the chart, and the axis of Mount Hedgehog, which runs nearly north and south, is placed on the chart running in an easterly and westerly direction.

After photographing the peak and planning a line of ascent, I left Trevor-Battye to sketch, and returned to camp to look for the men who were bringing our supper. I scanned the glacier in vain, and cursed my stupidity in allowing them to take down the whole of our sixty-feet rope, leaving us imprisoned on an island of rock, surrounded by névé crevasses, covered with new snow. As I watched for the men the fog gradually came up, and, before long, it settled down upon us as thick as ever. I realised that our brief chance of making the ascent was gone.

I was musing on the hollowness of life, when my attention was attracted by a moving object on the other side of the tent. At first I thought it was a bear, but presently discovered that it was our men returning from the opposite direction to that in which I was looking for them. In spite of my injunctions not to leave our track, they had tried a short cut, which had landed them in the midst of the maze of crevasses surrounding camp. Their appearance was decidedly ludicrous. Bottolfsen, who had made the acquaintance of the interior of five crevasses on his way up, was plastered with snow to the eyes; he was hung all over with an odd assortment of objects—cooking-tins, biscuit-boxes,

&c., like the White Knight in "Alice through the Looking-Glass." These things, he told me, had originally been contained in a blanket on his back, but they had strayed during his temporary visits below ground.

Summoning Trevor-Battye, who arrived shivering with cold, and with the paints frozen solid on his block, we set up the tedious operation of boiling snow for supper. About eleven P.M. we saw that it was hopeless to start that night, for the wind was rising, and it was snowing heavily, so we resigned ourselves to fate, and tried to sleep. Presently the wind increased to a gale; we had to turn out and lay big stones round the floor of the tent to keep it from being blown away bodily.

We looked out at five A.M.; it was still snowing, and nothing was visible but a great whiteness. At ten we cooked some coffee; the wind, which had moderated somewhat, again increased to a gale. Bottolfsen cheered us with Arctic yarns, and we inspected the weather at intervals. At 12.30 we performed our ablutions and had some lunch. The ablutions were scanty, owing to the absence of water, and chiefly consisted of a rub with snow—what the French would call *nettoiage à sec.* We took stock of our remaining provisions, not a difficult operation, as they consisted of one tin of Irish stew, a ration cartridge, a few biscuits, chocolate, and sticks of Emergency Food. The suggestion was made that, as soon as it cleared a little, we should start down for the ship; but I was loath to abandon the expedition as long as we had food, and stated my intention of sticking to it, and at least of climbing on to the foot of the mountain and obtaining specimens of the rock. Accordingly, at eight P.M. we put the remains of our provisions in our pockets, wrapped ourselves up in everything we possessed, and started for the peak, leaving the engineer buried under the tent, from which we had abstracted our ice-axes. Our last orders to him were that he should begin melting snow when he heard us shout.

At first we kept up the ridge from our camp-island to the wide saddle which separated us from the western buttress. By crossing exactly along the top of the col we avoided a gaping bergschrund, which circled round the head of the glacier basin. As the only mountaineer of the party, it fell to my lot to lead. Desirous of avoiding as much step-cutting as possible, I kept to the top of the arête, but my companions did not fancy the rocks, which were smooth and hard, so we descended to the edge of the snowfield falling from the South side of the ridge. We kept along this as far as the depression at the foot of the western buttress, by which I proposed to ascend. The fog cleared for a moment when we were on the col and opened a glimpse to the north over the head of Horn Sound. The sun was very near the horizon, glowing like a ball of molten metal, while the icebergs in the bay caught the glint of orange light and flashed it on to the surface of the water. The whole scene was vignetted in a fog, whose margin was lit up with a crimson glow, completing a most exquisite picture. It was the only view we had. During the rest of the climb we could scarcely see more than a few yards around.

When Trevor-Battye had made a rough sketch we commenced the ascent of the buttress, skirting always up the margin of the snowfield, which gradually narrowed to a steep couloir. Tired of kicking steps in the snow with boots which were too short for me, I again took to the rocks, but a growl or two from my companions sent me back to the snow, into the narrow part of the couloir, where steps became harder to kick and soon had to be cut. Gradually the surface of the old snow changed into névé-ice and finally to blue ice. In some places this was covered with as much as a foot of fresh snow, which had avalanched from the steep sides of the couloir and made step-cutting more difficult. At length we reached the foot of the final tooth, and with some difficulty gained a footing on the rocks. These were very smooth and plastered with ice, whilst the covering of fresh snow obliterated all hand-holds. After proceeding a short distance Trevor-Battye declared that he had had enough and proposed to await our return. I glanced at Bottolfsen, who said nought, but looked things unutterable.

Under suitable conditions of weather and snow, half-an-hour's climbing would probably have taken me to the top, but the mountain was in no state for solitary climbing, and the weather was as bad as could be. It was snowing again, and only a few feet of the base of the tower were visible through the fog. There was nothing to gain by continuing the ascent, so I read my aneroid, which gave the height as 4400 feet. Scoresby gives the height of the mountain as surveyed from the coast at 4395 feet, and that of the more northerly peak as 3306 feet, while the chart marks the summit at 4480 feet. We agreed at the time that the tower rose about 70 or 80 feet above the point where we stopped. I looked at my watch ; it was just 12.30 A.M.

Turning back, I found my companions had already begun to descend. When we had returned a short way I remembered a stick we had brought up to leave as a memento of our ascent. We had sacrificed one of our few remaining tent pegs for the purpose, squared it with a knife, and carved our names on three of the sides, and the date on the fourth. It was now deposited as safely as possible on the rocks and the descent continued.

All went well for a time, but the steps gradually became more difficult to find, so I changed places with Bottolfsen, who had hitherto been leading down. The position was an awkward one : Bottolfsen had no axe, but only the broken fragment of the pole he had brought up. I left him my axe afterwards and tried the pole, and I can only say that his performance, in coming last down the upper part of the couloir, was exceedingly creditable. Indeed the whole behaviour of my two companions was most plucky throughout the expedition. Neither of them had previously ascended a snow mountain, and their perseverance under the conditions in which we found the peak redounded greatly to their credit. I know that their chief reason for accompanying me was pure good nature. Not being mountain enthusiasts, they must have found the ascent tedious and trying in the extreme.

Coming down was not much more rapid work than ascending, though it was vastly easier for me; in the end, however, we reached the snow col a little before five o'clock. Here the fog was so thick that farther advance was fraught with considerable risk. The saddle had to be crossed along the top. Any deviation to one side or the other would inevitably have landed us in one of the bergschrunds which swept round the head of each snow-basin, nearly joining on the saddle. We had to hit off the narrow place between them. Not a landmark could be seen. The snow which had fallen during our ascent had completely obliterated our footsteps. After turning in every direction until we lost our bearings, we had to follow our tracks back to the rocks. No other resource being left, we set our compass in the general direction of Bastion Ridge, and plunged into the unknown.

After some minutes of considerable anxiety, expecting at each step to plunge headlong through the lip of a bergschrund, I noticed marks on the snow at my feet; stooping down to examine them I found that I was walking directly in our former tracks. Jodelling with all our might to the imprisoned engineer, we hurried down Bastion Ridge, and at 5.20 A.M. were under the friendly shelter of the tent.

Our Emergency Food, which we had been too miserable

to consume on the peak, was eaten and washed down with unsweetened coffee. Just as we had finished, Trevor-Battye, with an expression of intense surprise, drew from his pocket a fragment of the rind of an old Dutch cheese which he had bought on the ship in Advent Bay. This he generously divided between us, and then our thoughts turned to a smoke, but our tobacco was finished, and we had only one match left. Raking up the ashes in our pipes we struck our last lucifer in breathless silence and solemnly handed it round.

After four hours' sleep we started down to the coast. The fog had lifted slightly, and we had no difficulty in finding our way to the boat. We learned that the captain had almost given us up in despair, and that the steamer, which we had hoped to catch on its way to Norway, had sailed two days before, after waiting for five hours in the bay. As our provisions were practically exhausted, there was no use in remaining where we were. While they were getting up steam, we made a short boat expedition up the bay, then set out to cross the three hundred miles of Arctic Sea which separated us from Tromsö and civilisation.

The voyage was decidedly lively, though not really rough. Our little launch was swept by wave after wave, and we had frequently to slacken to half-speed to avoid diving underneath altogether. Matters in our little cabin reached a climax of discomfort when the water, which was rushing about on deck, poured down the companion in one stupendous leap (as they say in the advertisements of waterfalls), on to our cabin floor, converting the place into a swimming-bath.

The boat was rolling heavily at the time. Before we could oust the unwelcome intruder, an avalanche of miscellaneous commodities shot over into the water from the shelf above. At the same moment a table, on whose equilibrium we had hitherto implicitly relied, distributed its contents impartially into the bath. The collection swimming about was decidedly mixed—birds, biscuit-tins, bread, butter, boots, note-books, and onions, jostled each other at every lurch; while a tumbler of cod-liver oil, which Trevor-Battye had brought as a specific against colds, floated in amber globules on the surface of the water. In fishing the things out I came across my watch, which had already lost its glass during the step-cutting on the previous day. Its immersion in sea-water

THE "WINDWARD" AND THE "FRAM" AT TROMSÖ.

completely finished it; nothing would induce it to go another tick. In his book on Africa, Dr. Gregory speaks of the treatment he recommends for watches that refuse to go: he oils them, pats them, and finally sings hymns over them. Mine had been thoroughly soaked in cod-liver oil; I patted it vigorously; but, as Dr. Gregory was not present to sing hymns over it, the charm failed to work. After this we barricaded the door with our casks and sacks of coal, and so after three and a half days at sea we made Tromsö harbour.

Thus ended rather an adventurous expedition, which brought us back, after our enforced starvation, with the best appetites that we had ever enjoyed, enabling us to do ample justice to the banquet, on the evening of our arrival, given to Dr. Nansen, who had just returned in safety from his marvellous voyage.

CHAPTER XXV

SPITSBERGEN AS A SUMMER RESORT

THE relative accessibility of Spitsbergen rendered it certain to become, sooner or later, the goal of summer tourists. Holiday-makers are incited to travel by a variety of motives. Curiosity moves some; mere fatuous love of change operates on others; whilst I fear that the only reason for which many leave their homes is to be able to boast on their return that they have visited such and such places beyond the range of their friends. The ardent lover of Nature is impelled to become acquainted with all the moods of his great mistress. It is his joy to behold her in sunshine as in storm, in the glory of fertility as in the majesty of the desert; to pursue her into the fastnesses of the mountains or the breadth of the plains; to know her beneath her mantle of snow and ice, as well as in all her gorgeous pageantry of tropical exuberance. For him the mystery of Polar snows and the summer-long day must have a strong fascination. Knowing that Nature never and nowhere "did betray the heart that loves her," he will feel confident that there is a beauty of the Arctic regions as well worth knowing as that of any other part of this great terrestrial ball.

The time, therefore, was sure to come when it would occur to some caterer for the public entertainment that tourist steamers to Spitsbergen would pay. The pioneer in this enterprise was Captain W. Bade, formerly an officer in the German Navy. He took part in the North-German

Polar expedition of 1869, when the ship to which he was attached was smashed in the ice, and her crew lived for 237 days on an ice-floe, drifting down the east coast of Greenland. Captain Bade increased his Arctic experience by various whaling expeditions, and in 1891 he brought up a party of Würtemberg tourists to Spitsbergen. The experiment was so successful that he has repeated it every year since, on a continually increasing scale. In 1893 the Hamburg-American Company's great steamer *Columbia* brought a cargo of visitors to Advent Bay. In 1894 the Orient Company's steamer *Lusitania* followed her example, and others have succeeded.

Finally, in 1896, Spitsbergen may be said to have been formally annexed by the ubiquitous tripper; for, not only did the enterprising Vesteraalen Steamboat Company institute a weekly service of steamers running between Hammerfest and Advent Bay during the six summer-holiday weeks, but they were even bold enough to erect, on the site of an old Norwegian hut on Advent Point, a small wooden inn. I understand that in 1897 they propose to offer even greater facilities. In addition to the weekly steamer and the inn, they will have in Spitsbergen waters two small, properly built wooden steamers, to carry visitors to various points of interest; whilst the whole service will be under the direction of Captain Sverdrup, Nansen's well-known companion in Greenland and across the Polar Ocean.[1]

To the ordinary traveller Spitsbergen cannot fail to afford interesting experiences. If he goes up fairly early in the year he will probably meet with drift-ice on the sea between Bear Island and the South Cape or Point

[1] Persons desirous of information about this matter should communicate with the manager of the Vesteraalen Steamship Company, Mr. With, Stokmarknaes, Lofoten.

Lookout. Presently the famous Hedgehog Mount or Hornsunds Tind will come in view, towering above all neighbouring hills, and producing the impression of a giant mountain. More or less of the west coast will next be seen, with glaciers coming down from the inland ice to the margin of the sea. He will look into Horn Sound, and will in all probability be taken into Bell Sound, once the harbour of the English whalers. Schoonhoven (improperly but commonly called Recherche Bay) will doubtless be visited. There Arctic glaciers can be investigated close at hand, and even walked upon without difficulty. It was not far from this bay that, in 1630–31, a party of English whalers spent the winter in the blubber-boiling hut, having been accidentally left behind. They were the first men who ever lived through a whole year in Spitsbergen; the account of their adventures made a great sensation in its time and is still worth reading. After passing Bell Sound, the mouth of Ice Fjord is quickly reached. Beautiful, indeed, and highly characteristic is the scene on entering, with the fine mountains on either hand, the great glaciers coming down from the north, and the strange table-hills stretching away to the south. Along the front of these the steamer passes for a few hours before rounding into Advent Bay.

Of course, in Spitsbergen, as in Europe, much of the pleasure of travel depends upon the weather. The chances of sunshine at any given hour of the twenty-four in any day of July appear to be about even. In August the sky is more frequently overcast, in September yet more frequently. But the interest, as distinguished from the charm, of Arctic travel does not depend upon weather. Fogs and low-lying clouds are characteristic of the Arctic regions, and give rise to many most beautiful effects. The temperature on fine, clear, still days is like that of a warm English spring, but such days are not, of course, a majority. Usually the weather is cold, and warm clothing is a necessity, especially on board ship. Ordinary winter clothes suffice; but a good fur-lined overcoat will be found a blessing. Very convenient fur-lined pea-jackets, with holes in front for the hands, are sold cheaply at Bergen and Trondhjem.

Advent Bay is not a specially interesting centre for Spitsbergen. The energetic traveller will, however, be able to find congenial employment in the immediate neighbourhood. He can climb Mount Nordenskjöld, if there is a chance of a view, and thence look abroad over the whole interior of the island. A very vigorous walker might even make the still more profitable ascent of Fox Peak in an eighteen hours' walk. The hills on the west side of the bay have never been ascended, and there is an important valley leading into the midst of them and debouching on Ice Fjord just outside the entrance to Advent Bay, which has never been explored. Ordinary visitors, however, will do best to take advantage of one of the little steamers and make an excursion round Ice Fjord to Sassen Bay, Temple Mountain, Klaas Billen Bay, and Cape Thordsen. The ascent of any hill in these directions is better worth while than an ascent from Advent Point.

The most interesting part of Spitsbergen for a rapid view is, however, not Ice Fjord, but the west coast and bays north of it. Here the mountains are formed of the hardest and most ancient rocks; their forms are precipitous, and they are adorned with needle-pointed crests and summits, resembling the Aiguilles of Chamonix. Moreover, here glaciers are more numerous and grand than farther south. An expedition to the north-west corner of the island is therefore an essential part of any well-conceived visit to Spitsbergen, however brief. Historically, too, this is the most interesting part of these seas, for the bays and sounds at the north-west were the centre of the whaling industry in its flourishing days in the seventeenth century. Every point is the scene of some tragic event, some shipwreck or disastrous wintering.

From the mouth of Ice Fjord the northward way for large steamers lies outside Prince Charles's Foreland, a long partly submerged range of mountains of fine form, which have never been climbed, nor even properly mapped. Smaller boats can go up the more interesting narrow channel between these mountains and the mainland, peering into various secluded bays as they pass, and perhaps even looking into the beautiful English Bay, so well decribed by Lord Dufferin in "Letters from High Latitudes." Beyond the Foreland come the seven great glaciers or "Seven Icebergs," flowing down side by side from the inland ice to the sea. Then follows the "Pearl of Arctic Scenery," Magdalena Bay, alone worth a journey to behold. The narrow bay is enclosed by precipitous peaks and draped with glaciers. A little low promontory on its south shore contains the ruins of numerous graves. It was the English burying-place in whaling days. Beyond this bay are the craggy and snow-decked Danes and Amsterdam Islands, which shelter Dutch Bay from the western ocean. By either of two narrow entries the large secluded harbour may be gained, where the main body of the whaling fleet used to ride, and on whose shore was planted the Dutch summer settlement, Smeerenburg, of which scarcely a brick remains. Here the wildness of the scenery culminates, the rocks are all splintered by frost, snow frequently lies deep by the very margin of the sea even at midsummer, whilst, in many years, the ice-pack of the Arctic Ocean reaches down to the immediate neighbourhood, so that its nature and expansion may be estimated by any one who will row to some of the islands of Fair Haven (Cloven Cliff or the Outer Norway for choice) and scramble to the top.

Farther than this to east or north it is not likely that tourist steamers will often go, nor is it advisable that persons with home engagements should risk the adventure of a region in which it is always possible to be entrapped by the ice. All the bays and fjords of the west coast, however, may be visited in safety by a traveller having an extra week or two at his command, if the company should make needful provision of small excursion steamers, as, I am informed, they intend to do. A fair specimen of the Arctic world is thus thrown open to every intelligent person, and the horizon of every one's experience is thereby potentially widened.

The success of the Vesteraalen Company's experiment, of course, depends upon the popular support it receives at the hands of the touring public. The risk and the pecuniary profit is their affair. But there is another and a larger profit which the success of this venture will bring to science, and it is in the interest of this profit to science that I am doing what I can to further the purely commercial interests of the Company. Before the development of Switzerland as a holiday resort the Alps were visited by men of science; but no one will assert that the minute knowledge we now possess of the great Alpine range would have been attained if the playground of Europe had been located elsewhere. Scientific men have availed themselves of the facilities afforded to tourists, and tourists in their turn, being for the most part persons above the average in intelligence, have created a demand for the information which scientific men could supply. Thus the Alps have been surveyed as no other range of mountains in the world has been surveyed, and a public has been provided to take an interest in Alpine science, which but for them would scarcely have come into being. The same thing will happen in Spitsbergen if summer travellers can be persuaded to frequent it. A portion of Arctic land will be minutely studied and exactly surveyed, its changes watched and recorded from year to year, its phenomena patiently investigated, and its record maintained.

The interior of Spitsbergen is an almost unknown region. It is now to be brought within ten days of London, and opened for investigation to any person with a six weeks' holiday at command. The glaciers are unmapped, the peaks unclimbed, the valleys, for the most part, untraversed, the ice-sheet absolutely unexplored. Come, then, all ye "who live in houses and go to offices," and taste the delights of the unknown! Your chance is brief, for in a few years the hills of Spitsbergen will be even as the Alps, where there is no more a virgin peak for a man to conquer. In the remainder of this chapter I propose to indicate some of the work that awaits the man of adventure, and how he should be equipped to undertake it. The question of equipment shall be considered first.

A traveller who would explore the interior of Spitsbergen and climb its mountains must be prepared to carry all his equipment on his own back, or the backs of other members of his party. Hence everything taken must be as light as possible. Thin Willesden drill "Mummery" tents are best, with the ice-axes used for tent-poles. There should be one of these tents for every two, or at the outside three members of the party. For sleeping-bags the Norwegian reindeer-skin sacks, though most comfortable, are far too heavy and bulky. The traveller must be content with bags made of eider-down quilt. One kilo of eider-down will suffice for a bag. The down may be purchased for about thirty kroner a kilo, through Mr. Mack of Tromsö. It should be made up in England in a cover of woollen sateen of a kind you can get at the Army and Navy Stores, and doubtless elsewhere. It pays to carry a rifle and a few cartridges, for reindeer can generally be procured; but the rifle should not be heavier than can be helped. A Paradox is the best kind to take. A change of foot-gear is necessary, for boots and stockings will be wet through daily. A piece of thin rubber-sheeting should form part of each man's pack; it will serve to keep the sleeping-bag dry on the march, and for floor to the tent in camp. For food you must carry biscuits, concentrated soups and stews,[1] brick-tea, and the like. When reindeer are not likely to be forthcoming, suitable ration cartridges must be taken, such as those manufactured by the Bovril Company; they are filling, if not exactly appetising. No good light cooking apparatus exists. The best way is to carry an aluminium saucepan, and to boil it by burning beneath it pure spirits of wine in a small open pan or tray about three inches wide. If this pan be put on the bottom of an empty biscuit tin with some stones round it to support the saucepan, the tin will keep off the wind, and form a more efficient cooking apparatus than the bulky and cumbersome affairs made for travellers by people who have never been away from a town. With such an equipment as this it will be easy to make expeditions of four or five days' duration from the coast, where the party should have a whale-boat more elaborately stored with comforts. With this boat they can row or sail from one base to another, and the whole western part of the island, and,

[1] The Portable Food Company's consolidated stews in tin cylinders are the best things of the kind known to me.

in favourable seasons, much of the north, will be accessible to them.

Those who would undertake longer and more venturesome journeys over the great sheets of inland ice must equip themselves with a supply of Buckingham's Alpine Club rope, and with one or more suitable Nansen or other sledges. They will also do well to learn to shuffle about on *ski*. Not before the middle of July is the snow so far melted as to make such a journey pleasant; but in August, and even the first part of September, rapid progress can be made, without snow-shoes or *ski*, over the hard surface, and great distances covered with little labour. By means of a sledge, heavier equipment may be taken, and consequently supplies for a longer absence from any base. Arrangements could doubtless be easily made at Advent Bay for a small steamer to meet the party at particular places and times. Of this, however, it is as yet too soon to speak. For the present it will be best for any party to provide themselves with a whale-boat, and one or two Norwegian seamen to help them with it and to assist in carrying loads inland. Such men can be hired through the Vesteraalen Company or through Mr. Mack or Mr. Johannes Giaever of Tromsö. A rigid and detailed agreement should be made with them as to the loads they are to carry, their own equipment, their food, &c. A whale-boat can be hired through the same persons or through the Military Equipment Stores (7 Waterloo Place, London, S.W.), from whom Nansen sledges, *ski*, Mummery tents, and so forth can be bought. Fur goods are best purchased from Mr. J. N. Bruun of Trondhjem, who sells also the useful fur-lined pea-jackets above referred to. Silver & Company of Cornhill supply all sorts of travellers' requisites, which I have always found well and strongly made. Edgington of London Bridge makes excellent tents.

Good equipment can also be bought at the Army and Navy Stores; but many things may be bought in Norway as well as in London, and for a less price. Very thick woollen clothes and underclothes should of course be taken, though Spitsbergen affords little opportunity for changing.

We come now to consider what opportunities for exploration Spitsbergen offers to a summer visitor. The country may be divided and considered under four categories—the

INLAND ICE-SHEET OF NEW FRIESLAND FROM HINLOOPEN STRAIT.

coasts, the mountains, the inland ice-sheets, and the snow-free valleys. The coasts have all been explored, and nothing remains to be done with them except by really expert surveyors and experienced men of science. The coast mountains, however, are little known, and only the easy ones have been climbed. I hope next year to publish a full account of what has been done in Spitsbergen in the way of mountaineering, and then any one can see for himself what remains to be done. The best climbing is to be found near the sea along all the west coast north of Ice Fjord. The Foreland Mountains are probably as high as any in this region, and one of them may even be higher than Hedgehog Mountain. They are all unclimbed, and even unmapped, save in the crudest manner. In the neighbourhood of Magdalena Bay there is also excellent climbing of steep and difficult rocks to be found. There are plenty of hills farther east, approachable from the fjords; but they are, I believe, in no case difficult, owing to the nature of the materials of which they consist.

The exploration of the snow-free valleys is scarcely begun. Any one who would devote a season to mapping and traversing the area westward of Advent Bay, between Ice Fjord, Bell Sound, and the sea, would be doing excellent work, for the region is most interesting both to geologists and physical geographers. There is a large snow-free area of uncertain breadth extending from Ekman Bay (of North Fjord) eastward to Klaas Billen Bay, and from Ice Fjord northward to Wijde Bay and along its west shore. Much good work might be done here from the moving base of a boat, especially rather late in the season in years when Dickson Bay is clear of ice. The third more or less snow-free region in which much remains to be discovered is that between the wrongly-named Van Keulen Bay (it should be called Sardam or Michel Rynier Bay, I believe) and the southern limit of the area explored by us. The most interesting part of this region is that penetrated by the deep valleys at the head of the two branches of Low Sound (misnamed Van Mijen on the chart). By either of these valleys, and especially by the southern, a rush might be made to the east coast.

The inland ice-sheets may be taken piecemeal. From Horn Sund and the head of Van Keulen Bay the whole southern sheet can be explored; but Van Keulen is seldom open till late in the season. The sheet north of the Sassendal can be entered by way of the Post Glacier in Temple Bay, or the Nordenskjöld Glacier at the north-east angle of Klaas Billen Bay. The Nordenskjöld Glacier is also the natural point of departure for the great ice-sheet stretching away north and north-east. There is a great glacier valley debouching on Hinloopen Strait almost opposite Cape Torell, which it would be most interesting to descend. The best glacial area of all to be explored in a short summer expedition is that between Kings Bay, the sea, and Ice Fjord, for it is traversed by splendid rivers of ice from south-east to north-west, and it contains a number of fine peaks which might be climbed. At the head of Cross Bay there is said to be a long snowy valley that runs northward; this and all the snowy and mountainous area north and east is full of opportunity for interesting discoveries.

In half-a-dozen summer holidays of six weeks' duration or thereabouts, an active party of men, familiar with ice-craft, and possessing such rudimentary knowledge of surveying as may be learnt from Mr. Coles at the Royal Geographical Society (1 Savile Row) in a few hours of study, might fill in the main outlines of the as yet blank interior of Spitsbergen, to their own great enjoyment and the profit of geography. The whole country is intensely interesting from a scientific point of view, because of the rapidity with which its surface is being modelled into such forms as were impressed in glacial times on the now temperate and inhabited parts of Northern Europe. For this reason we need accurate and detailed information about the nature of the interior of Spitsbergen, as well as records from year to year of the changes that occur in it, so that we may be able to deduce the rate at which valleys eat their way

SKETCH MAP
OF PART OF
SPITSBERGEN
From a survey by Sir W. Martin Conway
in June, July, and August, 1896.
SCALE OF MILES
0
10
Heights in feet.
Inland ice
WICHE BAY
S T O R
O R D
Robert Glacier
DUNER BAY
Ulve Glacier
AGARDHS BAY
FOX NOSE
CAPE DUFFERIN
STOR FIORD
or
WIJBE JANS WATER
Ingelfield Glacier
BAY

back, mountains are cut out of table-lands, bays are filled up, and so forth. Here, then, is a chance for competent men to enjoy holidays of an active, health-giving, and novel sort, and at the same time to perform good and fruitful service to science.

FAREWELL!

APPENDIX

THE NOMENCLATURE OF SPITSBERGEN

SPITSBERGEN was discovered and named by the Dutch in 1596. They thought it might be part of Greenland. The name Spitsbergen was only given to the land with the pointed peaks, and does not belong to the whole Archipelago. North-East-Land, Barendsz Land, and Edge Land are not parts of Spitsbergen. The English used to call Spitsbergen Greenland, till a relatively recent date; the same name is also found, perhaps more frequently than Spitsbergen, in foreign authors. Barendsz only named two points on Spitsbergen—Vogelhoeck (the N. point of the Foreland) and Keerwyk (the N. end of Foreland Sound). The other three names on his map—Gebroocken Land (the islands at the NW. angle), Grooten Inwyck (Ice Fjord), and Inwyck (Bell Sound)—are descriptions rather than names.

The next visitor to Spitsbergen was the English skipper Hudson (1607), who called the island New Land, or King James his New Land. He had Barendsz' map with him, and did not attempt to change the Dutch names, but he added a few of his own—Collins Cape (probably the Hakluyt Headland of to-day), Hakluyt's Headland (probably the N. cape of North-East-Land), Whales Bay (?), and a few more. Other English skippers gave other names to new points. Then came Dutch, German, and French whalers and added names of their own, changing the rightful designations, and often misapplying old names, so that the same name is found in different charts applied to different points, and different names to the same point. Finally the Norwegian skippers and the scientific explorers of the present century have created further confusion, so that now the nomenclature of this region is in a condition of astonishing confusion. I have, therefore, made a careful study of all the old MS. and other charts to which it has been possible for me to obtain access, with a view to discovering what was the original designation of each point—a matter of no little importance and interest, seeing how much local history is fossilised in names. My researches are far from complete. The present chapter is a mere sketch, which will doubtless need revision hereafter. I propose to start at the south Cape, and work up the west coast of Spitsbergen and round the whole island, taking the various named points in geographical succession. I shall then similarly treat the neighbouring islands. The references are to the various authorities named at the close of the chapter. I have not included designations from charts or books later than Scoresby (1820). References to Martens are to the following book :—F. Martens, *Spitzbergische Reisebeschreibung*, Hamburg, 1675 (English translation, Hakluyt Society, 1855).

Ronde Klip (Middelhoven, 1634).—The island off the S. Cape.

Point Look-Out.—So named in Hessel Gerritsz (1613). In Dutch it is *Kaap de Kyckuyt* or *Uyt Kyk* (*passim*); *Generaels hoeck* (Joris, 1614). *Première pointe* (Cash, 1629); *Whale's Back* (Van Keulen, 1689).

Mount Hedgehog.—This is the name given in Purchas to a prominent peak at the south end of the island, which can be none other than the *Horn Sunds Tind* of the Norwegians. There is only one mountain in this part, visible from both coasts, and from far to the south, which by far overtops all neighbouring peaks. It is called *Mount Edge* in Edge's chart (1625); misprinted *Mound Egle*, and misplaced by Van Keulen (1689); Gerritsz (1613) calls it *Moscovit Mont*, but this is probably the peak on the E. coast so named in Edge's chart, which appears as *Mount Hedgehog* in modern maps. Scoresby is the first authority I have yet found for the name *Horn Mount*.

Horn Sound.—Named by Hudson (1607). The name is confused into "*Horv Sond* or *Hoorn Sond*" (G. and R., 1707); and into *Horizond Bay* (Zorgdrager).

Dun Islands.—These appear to be the *Lammas I.* of Hondius (1611).

Rheeland (*Roebuckland*), named first in Van Keulen (1689).

Slaadberg (G. and R., 1707), a hill between Schoonhoven and Dunder Bay.

Bell Point.—The promontory S. of Bell Sound, so named in Gerritsz (1613) and ever since.

Bell Sound.—The *Inwyck* of Barendsz. The name first appears in Joris (1614). It is the designation of the whole group of sounds or merely of the mouth outside the junction of the inner bays. Sometimes it is called *Clock* (*Klok*) *Bay* or *Rivier* (first by Blaeu 1664). The nomenclature of the different branches of this Sound is the most confused in the whole island. It divides into three branches, (*a*) *Schoonhoven* (Recherche Bay) which goes S.; (*b*) *Sardam Bay* (Van Keulen) which goes SE.; and (*c*) *Low Sound* (Van Mijen), which goes NE. The difficulty of finding the correct nomenclature arises from the fact that many of the early charts depict four branches, the fourth being a duplication of Sardam Bay as a branch of Schoonhoven. Taking the branches in order they are:—

Schoonhoven.—So first named in Gerritsz (1613), and continuously thereafter (or in English *Fair Haven*, *Clean Bay*, or merely Bell Sound). It was the great place of assembly for the whaling fleet about to return home, during two centuries. The alteration of this name to *Recherche Bay* by the French Expedition, which spent eleven days there in 1838, was a most unwarrantable mutilation of history.

Rheen Eylandt (Blaeu, 1664).—An island at the head of Schoonhoven.

Sardam Bay (Guérard, 1628).—Described in Gerritsz (1613) as *Baye des Franchoys.* The name *Michel Rinders Rivier* also appears in Guérard in connection with this bay, so that one name or the other may apply to some minor bay within the larger. The name of Van Keulen in connection with these waters first appears as *Van*

Keulens Baaytje (G. and R., 1707) or little bay, evidently applying, not to the whole Sound, but to some sheltered anchorage, probably the enclosed inlet at the extremity; Keilhau (1827) calls this *Mittel-fjord.*

English Tent (*Plaats van't Engelshuysje*, G. and R. 1707).—Situated either on or just within the Pt. Ahlstrand of the chart.

Point Partition (Gerritsz, 1613).—Between Sardam and Low Sounds.

Low Sound (Gerritsz, 1613).—The original English name. *R. de Kloeck*, or *Klok Bay* or *Rivier* appears first in Guérard (1628), and becomes the regular Dutch name (except in Blaeu's text). In some of the preceding chapters I have wrongly used the name Klok Bay for Low Sound. It is wrongly called *Rinder's Bay* in some Dutch charts (Commelin, Van Keulen). Recently it has been generally known as *Van Mijen Bay*, but this is the mutilated name of a neighbouring anchorage.

Coude Herberch (Commelin, 1624).—A name frequently given in old Dutch charts to some anchorage near the head of Low Sound.

Ondiepe Rivier (Blaeu, 1664).—Described by G. and R. (1707) as "Een drooge Fioerd vol Mouras dar sig veel Rheenen outhouden, genaamt Ondiepe Rivier." *Drooge Riv.* (Zorgdrager), *Dry Fiord* (Scoresby). The Norwegian reindeer hunters call it *Stordal.*

Willem van Muyden's Haven.—A small bay in the N. coast of Bell Sound, west of Axel Island. It was named after the first Dutch interloping skipper who sailed for the Spitsbergen whaling. The bay is called *Bottle Cove* by Pellham ("God's Power," &c., 1631), *Fair Haven* by Scoresby.

Lowsoundness (cape N. of Bell Sound).—So named in Gerritsz (1613). *Lowenesse* (Edge, 1625) was probably the original name, and Low Sound was named from it.

Lisetts Islands.—Named in Edge (1625) and *passim* till Scoresby (Lissetts). They were placed off the coast between Bell Sound and Ice Sound, where no islands are now marked.

Ice Sound (the *Grooten Inwyck* of Barentsz).—First named in Gerritsz (1613). Hudson (translating Barendsz) wrote of it as "the great Indraught," or perhaps "the Sack." It is named *Baye des Panoles* in Cash (1629). Sometimes it is wrongly called *Groenhaven* (Blaeu, 1664).

Russe Keglen (Norwegian name).—The valley between Green Harbour and the W. coast.

Green Harbour (Gerritsz, 1613).—A name that has never been changed.

Coles Bay (Pellham, 1631).—Of late years generally but wrongly called *Coal Bay*.

Advent Bay.—A modern name. The true old name was *Klaas Billen Bay* (G. and R. 1707), but this name has been transplanted across the Sound, and any attempt to alter it now would lead to hopeless confusion.

Sassen or *Sassele Bay* (G. and R. 1707).—Zorgdrager marks Sassen Bay as the name of the whole inner part of Ice Sound.

Gips Hook was called *'t Middelland* by the Dutch.

Safe (*Behouden*) *Haven* (Gerritsz, 1613).—A name that has seldom been confused. This bay was called *Niches Cove* by Edge (1625).

Osborne's Inlet (Gerritsz, 1613).—It is named *S. Jans Bay* by Middelhoven (1634) and all later cartographers.

Cove Comfortlesse (Gerritsz, 1613).—The *English Bay* of all later writers.

Prince Charles Foreland or *Kijn Island* (Gerritsz, 1613).—Generally named *the Foreland* in old charts. *'T lang Eylandt* (Blaeu, 1664).

Saddle Point.—The S. point of the Foreland. It is possible that when first seen, the hill at the S. end of the Foreland was an island separated from the rest. Even now it has that appearance except from close at hand, when the low intervening land appears. There are two names—*Black Point* (Gerritsz, 1613), *Kyn's Cape, Kianas* (Goos, 1666), or *Kynnae* (Commelin, 1642), applied to this South Cape or *Zuydhoeck* (Goos, 1666); but, of these names, the former at any rate belonged properly to a point on the main mass of the island.

Saal Berg (G. and R. 1707).—The hill behind Saddle Point.

Black Point (*Swartenhoeck*) (Gerritsz, 1613).—The point at the S. end of the main hill-range of the Foreland. It is called also *Middle Hook* by Scoresby.

Persch Riff.—A shoal perhaps represented by the problematical *Goshawk Rock* of the chart. Just S. of it there is written on G. and R. (1707), "Hier begint 't Gobergti of de Zuyd Hook."

Cape Siettoe (Colom, 1648).—A point on the W. coast of the Foreland, N. of Black Point.

Cape Cold.—Named in Gerritsz (1613), perhaps the same as Cape Siettoe.

Vogelhoeck (Barendsz, 1596).—The N. point of the Foreland. It was called *Faire Foreland* by the first English skippers, and sometimes the *North Hook*.

Pieter Winter's Baaytje (G. and R., 1707).—A bay in the E. coast of the Foreland near its S. end. A creek in the N. shore of this bay is called *Zeehonde Bay* by Zorgdrager and others.

The Barr (Edge, 1625), or *'t Riff* (G. and R., 1707).—The shoal that nearly joins the Foreland to Spitsbergen. Here Barendsz, coming from the N., had to turn back.

Foreland Sound (Van Keulen, 1689).—Between the Foreland and Spitsbergen. The part of this, N. of the Barr, was called by Barendsz *Keerwyk;* by the early English whalers, *Sir Thomas Smith's Sound* (Joris, 1614); by Scoresby, *Bay of Birds.* Zorgdrager writes of Foreland Sound, under the name *Hinter Foreland.*

Quade or *Kwaade Hoek* (Van Keulen, 1689).—The low spit S. of Kings Bay.

Kings Bay (*Koninks Bay*, G. and R., 1707).—Has had a great many names. It was called *Dere Sound* by Gerritsz (1613), a name given by Scoresby to the NE. harbour of Kings Bay. Other names are *Pt. des Gars* (Guérard, 1628), *Kar Soudt* (Middelhoven, 1634, and others), *English Bay* (Cash, 1629, Zorgdrager and others).

Three Crowns.—Marked but not named by G. and R. (1707).

Cross Bay or *Road* (Edge, 1625), called *Closse Sound* in Gerritsz, but this may be a misprint. The name has been constantly employed.

Mitre Cape (Scoresby) or *Cape Mitra*, apparently not an old name.

The Seven Icebergs.—So named in Zorgdrager.

Hamburger Bay (G. and R., 1707).—Called *S. Jans Bay* by Zorgdrager.

Little Basque Bay (Guérard, 1628).—A small bay between Hamburger Bay and Magdalena Bay, not marked on the Admiralty chart. It is named *Port Louis, ou Refuge François* in Cash (1629).

Magdalena Hook (G. and R., 1707).—The point S. of the entrance to Magdalena Bay, named Whale's Back on the Chart.

Magdalena Bay (Joris, 1614).—This may be the *Teeth Bay* of Barendsz' log. Scoresby names the SE. harbour within the bay, *John Duncan's Bight.*

Knottie Point (Gerritsz, 1613).—The point between Magdalena Bay and South Gat.

Third or Niew Makkelyk Oud.—This is marked on the S. coast of the S. Gat by Zorgdrager and G. and R. (1707). The latter also here marks *Ian Donker 't Varken Sonder heost.*

South Gat (Blaeu, 1664).—It is also called *English Bay* by Zorgdrager and many more, but the true position of this English Bay was really at the SE. corner of Dutch Bay.

Danes Island (Goos, 1666, and all later cartographers).—Its SW. point was called *Engelsche Uyt Kyck* by the Dutch (*passim*). The bay in its W. coast is marked on the Cash map (1629) *Port St. Pierre apellé par les danois Copenhavre baie.* Its common name was *Robbe* (*Kobbe*) *Bay* (Middelhoven, 1634), or *Danes Bay* (Zorgdrager). The small bay in the N. coast of this island is known as *Pike's Bay,* since Mr. Arnold Pike built a house there apparently on the site of the old *Kokery of Harlingen* (wrongly called Haarlem by Martens), behind which was "the running water" of Martens (*Vars water* of G. and R., 1707).

Danes Gat.—Perhaps oftener called *Middelgadt* (Middelhoven, 1634), by the Dutch, and frequently *South Bay* (Zorgdrager) with reference to Smeerenburg. *Dodmans Eyland* was at the E. end of Danes Gat. Zorgdrager marks *Zetje Fau* in or about Danes Gat, but I know not to what the name refers.

Amsterdam Island (Blaeu, 1664).—The chief hill on this island is named the *Beehive* (Martens). The north point, now called *Hakluyt's Headland,* was the *Kwaade* or *Devil's Hook* of the Dutch, and was probably the *Collins Cape* of Hudson (misprinted *Colnis* by Hondius, 1611). In Gerritsz this point is named *Ysse Caep* (1613). A small island off the point was named *Devil's Island* (Van Keulen), or *Bikorf* (Zorgdrager), and near it was the *North Bank* (Zorgdrager). On the flat SE. spit of Amsterdam Island was situated *Smeerenburg,* with five principal tents, cookeries, or settlements, named, in order from N. to S., *Amsterdam, Middelburg, the Danes, Delft,* and *Hoorn.*

Dutch or *Mauritius Bay* (Joris, 1614).—The name of the chief harbour in Spitsbergen, between the mainland and the Amsterdam and Danes Islands. It is the *Beerhaven* of Commelin (1642), and the *South Haven* of Martens. Its N. entrance was called *North Gat* (Blaeu, 1664). In its E. coast were reckoned to be three bays (Van Keulen, 1689): *English* or *South Bay* at the SE. angle; *Beere Bay,* farther N.; and *Slaad Bay,* still farther N. *Ys hoek* separated Beere Bay from Slaad Bay. Zorgdrager and G. and R., (1707) mark a river (*Zuyd Bay Rivier*), rising from a lake in the mountains to the S., and flowing into English Bay. This river and the minor bays seem to have been altered or filled up by glacial changes and encroachments. Zorgdrager marks a *Makkelky*

Oud on the E. shore of the bay, as well as a point named *Wage Padt*.

Om den Oost (Zorgdrager).—A name given by the Dutch to the N. coast, and the sea off it.

Large Point (Blaeu, 1664).—The *Foul Point* of the Admiralty Chart.

Fair Haven (Gerritsz, 1613).—The harbour enclosed between Vogelsang, Cloven Cliff, the Norways, and the mainland. It was also named *Hollantsche Haven* (Joris, 1614); *North Haven* (Martens); *Somer Bay* and *Bay met de Eylanden* (Blaeu, 1664).

Vogelsang (the old Dutch name).—Also named *Cape Barren* by Edge (1625), and *Rotganse Eyl.* by Blaeu (1664).

Cloven Cliff (old Dutch name).—The *Saddle Island* of Edge (1625); *De Reus* or *Giant* of Joris (1614); *Klip met de Kloff* of Van Keulen (1689).

Outer Norway Island.—This is the *Beare Island* of Edge (1625); the *Gansen Eyland* of Van Keulen (1689) and others; and the *Half-Norwegen* of Zorgdrager.

Inner Norway Island was called *Zeeusche Uytkyk* by Van Keulen and Zorgdrager. A shoal off Flat Hook is named by Zorgdrager *Rift van de Uytkyk.*

Birds Bay, the *Foul Bay* of the chart, but this name does not occur in any old map. It is marked *Grote Bay* by Joris (1614); *Bay aux oiseaux* by Guérard (1628); *De grote vogel bay* by Commelin (1642); *Archipelago* by G. and R. (1707) and Zorgdrager. Perhaps this is the *Vausques* (Basques) *baie* of the Cash chart (1629).

Fox Point (Edge, 1625).—The Dutch used to call it *Flat Hook* (*Vlacke Hoek*).

Monniers Bay (Guérard, 1628, and all the early charts); Commelin (1648) calls it *S. Laurens Bay*. The modern name, *Red Bay*, first appears with Van Keulen (1689). At the bottom of the bay was *Point Deceit* (Edge, 1625). The cove W. of Point Deceit is named *Ayer Bay* on Blaeu's map (1664). Thomas Ayers was one of the first English whaling skippers at the beginning of the 17th century.

Biscayers Hook (G. and R., 1707).—This was called by Edge (1625) *Point Wellcome*, but that name has now been shifted, past correction, to a neighbouring headland.

Red Beach (Edge, 1625) and *Rehenfelt* (Martens).—These are names for the land between Biscayers Hook and Welcome Point. The east side of the same cape, S. of Welcome Point, is named *Agter*

Reene Velt (G. and R., 1707, and Zorgdrager). G. and R. mark three hills on Red Beach—*Rooberg*, *Trourenberg*, and *Berg op Reenvehl*. Martens (1671) says that there is a hill on Red Beach "that looketh like fire."

Welcome Point.—The modern name for the point W. of the entrance to Liefde Bay. The original Welcome Point is that now known as Biscayers Hook.

Wiches Sound (Edge, 1625), *De Oostinwyck* (Blaeu, 1664), *Oosterwyk* or *Wyde Bay* (Van Keulen, 1689), *Liefde Bay* (Martens, 1671). The S. branch of Liefde Bay is named *Wood Bay* by Scoresby. A cove in the E. shore of Liefde Bay is called *Muys Haven* by Blaeu (1664) and *Liefde Baytje* by G. and R. (1707).

Moffen Island (Blaeu, 1664).—This island is marked in all the old Dutch charts. G. and R. (1707) state that it "is 6 veet boven water," so that Phipps was wrong in thinking it newly elevated.

Castlins Point (Edge, 1625), between Wiches Sound and Wijde Bay. It has had many names—*Swartehoek* (Blaeu, 1664), *Grey Hook* or *Flakke Point* (Van Keulen, 1689), *Derre Hoek* (G. and R., 1707), *Dorre Hoek* (Zorgdrager). This is the *Grey Hook* of the Admiralty Chart.

Wijde Bay (Goos, 1666).—Originally named *Sir Thomas Smith's Inlet* by Edge (1625). In its W. side, near the entrance, was a little bay named *Jan Tennises Bight* by G. and R. (1707), apparently the same as the *Beere Bay* of Van Keulen (1689). Farther in, on the same side, four reefs were marked by G. and R. (1707). On the E. shore G. and R. mark *Sand Dunes*, and farther out three great glaciers. *Aldert Dirkses Groote Baaytje* is marked by G. and R., whilst the point immediately N. of it seems to be the *Steyle Hoek* of Van Keulen (1689). *Bangen Hoek* (Van Keulen, 1689) is the point W. of Mossel Bay.

Mossel Bay (Martens, 1671) was named after Mossel, who was Van Muyden's second in command in 1612 or 1613. *Helvemaens Bay* (Goos, 1666) was an alternative name marked on most Dutch charts.

Verlegen Hook (Van Keulen, 1689) appears to be the *Point Desir* of Edge (1625). It is the *Langenes* of Blaeu (1664), and the *Vlakke Hoek* of G. and R. (1707). The name is misprinted *Vertegen Hook* on some modern maps.

Hinloopen Strait (Blaeu, 1664).—Also called *Weighatt* by the Dutch. Both names occur on most charts.

New Friesland (G. and R., 1707).—The country between Wijde Bay and Hinloopen Straits.

Willem Volckx Baaiytje (G. and R., 1707).—This is either Treurenberg Bay or some cove between it and Verlegen Hook.

Luysen Eyland (Van Keulen, 1689, and G. and R., 1707).—Marked as near the coast between Verlegen Hook and Treurenberg Bay.

Treurenberg (G. and R., 1707).—A hill W. of Treurenberg Bay.

Treurenberg or *Sorge Bay* (Scoresby).—Named also *Mosiel Bay* erroneously by Blaeu (1664). *Beere Bay* (Goos, 1666, and many others) seems to have been its common Dutch name. A hill S. of the bay was named *Beere Berg* (G. and R., 1707), but the same map marks *Lommeberg* or *Beere Berg* W. of Lomme Bay, and *Lomme Berg* again E. of Lomme Bay.

Parrot Hook (G. and R., 1707), between Cape Foster and Lomme Bay By "parrot" is meant "puffin."

Lomme Bay, or *Bear Bay* (G. and R., 1707).

Lomme Berg (G. and R., 1707).—E. of Lomme Bay.

Duym Point (G. and R., 1707).—As on the modern charts.

South Waygat (Scoresby).—The S. end of Hinloopen Strait.

Unicorn Bay (Goos, 1666), *'t schip d' Enhoorens bay.*—G. and R. (1707) call it *Lonie Bay*, or *de Eenhoorens Bay*.

Heley Sound.—Named after a fellow-skipper by Edge (1625); it is generally misspelt *Helis Sound*, or even *Hell Sound*. Poor Heley!

Seal Island (Van Keulen, 1689).—This is overwhelmed by a glacier, and is probably the hill now called *Edlunds Mount*.

Walrus Island (Van Keulen, 1689).—West of Seal Island, and like it now enveloped in glacier.

Bear Gat (Van Keulen, 1689).—Buried beneath the Negri Glacier.

Wiches Bay (Edge, 1625).—Doubtless much covered by the Negri Glacier; in the Dutch charts it is generally called *Whales Wiches Bay*.

Keerweer (Van Keulen, 1689).—The *Mohn Bay* of modern maps.

Fox Nose (Edge, 1625).—Or *Ness*.

Foul Sound (Edge, 1625).—The name is retained in all the Dutch charts except Blaeu's (1664), where this is called *Bay met Vuijlerudsen*. The modern name is *Agardh Bay*.

Whales Head (Edge, 1625).—This name has never been changed.

Whales Bay.—Not marked in Scoresby or any earlier chart.

Muscovie Mount (Edge, 1625).—Name retained on most old Dutch charts, but changed to *Mt. Hedgehog* on modern maps.

't Hol van een Schip (Blaeu, 1664).—An island or bay off the E. coast near to the S. Cape.

Wybe Jans Water (Middelhoven, 1634).—Amusingly called by Guérard (1628) *Destroict de Jean Suatre.* The Norwegians call it *Stor Fjord.*

BARENDSZ ISLAND

When this island was believed to join Spitsbergen it was called *South-East-Land* (G. and R., 1707), a name that continued in use till after Scoresby.

Lommenberg (Goos, 1666).—A hill on the E. coast.

Vossen Bay (Van Keulen, 1689).—In the W. coast.

Anderson's Islands were originally named by Edge (1625) *Sir Thomas Smith's Islands,* a name now generally applied to some smaller islands in the W. entrance of Alderman Freeman's Inlet.

Cape Barkham (Edge, 1625).

Alderman Freeman's Inlet (Edge, 1625).—Called *Walter Thymens Fioerd* in Van Keulen's chart (1689), and later.

EDGE ISLAND

So named by Thomas Edge in 1616. Its E. cape was named Stones Foreland, which was confused by the Dutch into *Stans Foreland* (G. and R., 1707) and applied to the whole island; this name is now commonly used by the Norwegians.

Lees Foreland (Edge, 1625).—The NW. cape.

Cape Blanck.—The N. point of Disco Bay.

Dusco (Goos, 1666).—A position on the W. coast.

Whales Head (Edge, 1625).—The SW. cape.

Deicrow Sound (Edge, 1625).—The deep bay between the SW. and S capes. It is named *London Bay* by Middelhoven (1634), and *Deeve Bay* by Goos (1666) and others. Van Keulen (1689) names the outer part *Deeve Bay,* and the head *Deicrows Sound,* and this is the nomenclature followed afterwards. *Bear Bay* (Edge, 1625) is in the E. side of the bay, close to Negro Point.

Negro Point (Edge, 1625).—The S. cape, called *Dictus Point* by Middelhoven (1634), *Swarten Hoeck* by Blaeu (1664), and *Black Hook* by many.

St. Jacobs Bay (Blaeu).—In the SE. coast.

Stans Foreland (Van Keulen, 1689).—A mistaken name for the SE. coast.

Disco (G. and R., 1707).—On Stones Foreland, near a former whale fishery.

Stones Foreland (Edge, 1625).—The E. cape.

Ryk Yse Islands (Goos, 1666).—Named after the Dutch skipper Ryk Yse.

Halfmoon Island (Goos, 1666).

Thousand Islands, a modern name, first found on Scoresby's chart, and suggested by the old maps rather than by nature, for the islands themselves are not really very numerous. Their number was greatly exaggerated on the Dutch charts, from Van Keulen's (1689). They are generally described as *Laag gebrooken Land.* The following names appear in the Dutch and other charts, but the islands cannot be certainly identified: *Abbots Island, Scott's Island, Heling Island, Hopeless Islands, Bear Island, Wester Island.*

Hope Island (Edge, 1625).

THE SEVEN ISLANDS AND NORTH-EAST-LAND

North-East-Land is marked on the Goos chart (1666) as a collection of islands, whereof the NE. group are called the Seven Islands. Van Keulen (1689) is not much better informed, but he marks two of the true Seven Islands separately to the north. North-East-Land is first properly marked and named by G. and R. (1707). The south point of it was seen by Edge in 1616, and named *Sir Thomas Smith's Island.*

The Seven Islands were well enough known in Martens' time (1671), but were first marked with approximate correctness by G. and R. (1707). *Hooyberg* (Van Keulen, 1689) was probably *Phipps Island. Tafelberg* (Van Keulen, 1689) was *Martens Island* or *Parry's Island. Ambeelt* (G. and R., 1707) may have been *Martens Island.*

Shoal Point (Martens).—The NW. cape of North-East-Land.

Great Stone (G. and R., 1707).—In its SE. side is marked a deep bay named *Illuys Haven.*

Low Island.—Perhaps this is the *Purchas plus ultra Island* of Edge (1625); it is *'t Lage Eyland* of G. and R. (1707).

Brandywine Bay (Van Keulen, 1689).

Extreme Hook. Marked by G. and R. (1707) *Uyterste Hoek* or *Uyterste Land.*

Point Purchas (Edge, 1625).—The *Black Cape* of Scoresby, the *North Cape* of modern charts.

Scoresby Island is apparently the *Vlak Eyl.* of G. and R. (1707).

Cape Platen seems to be the *Rene Eyl.* of G. and R. (1707).

Outger Reps Eyland (G. and R., 1707).

Walrus Eyland (G. and R., 1707) is Zoogdrager's *Walvis Eyland.*

Duyre Bay (G. and R., 1707).—SSE. of Outger Reps Eyland.

Een Groot hoog Eyland (G. and R., 1707).—Near Cape Leigh Smith.

Cape Torell.—A modern name for the point that used to be called *South Hook* by the Dutch (G. and R., 1707).

Veene (G. and R., 1707).—Apparently the modern *Augusta Bay.*

Wiches Land (Edge, 1625).—This, I think, is the same as that marked by G. and R. (1707), "Commandeur Giles Land ontdet 1707, is hoog land." In a future work I shall state my reasons for this belief. Wiches Land is now generally called *King Karl's Land,* or *King Charles Islands,* but the old name should be preserved.

LIST OF EARLY MAPS OF SPITSBERGEN

The following is not intended to be a complete list of maps of Spitsbergen. It contains only those which present novelties of nomenclature. Other maps and charts, which merely copy the names appearing on these, are not mentioned, nor is any map included that is later than Scoresby's :—

Barendsz' map, dated 1598, and published by Cornelius Claeszoon at Amsterdam. It is found in some copies of the second part of the abridged Latin edition of Lindschoten's *Navigatio ac Itinerarium* (The Hague, 1599).

Jodocus Hondius' map, published in J. I. Pontanus, *Rerum et Urbis Amstelodamensium Historia* (Amsterdam, 1611, 4to).

Hessel Gerritsz' map, copied from the lost chart by Daniels. It forms part of Gerritsz' *Histoire du pays nomme Spitsberghe* (Amsterdam, 1613, 4to).

Carl Joris' map of 1614; MS. in the Dépôt de la Marine, Paris.

Thomas Edge's chart of 1625, in the third part of "Purchas his Pilgrims," and in Pellham's "God's Power and Providence" (London, 1631).

Jean Guérard's chart of 1628, in the Dépôt de la Marine, Paris.

Cash chart of about 1629, French MS. chart belonging to Mr. C. G. Cash of Edinburgh.

Middelhoven's chart of 1634, MS. in the Rijks-Archief at the Hague.

Isaac Commelijn's map, based on Daniel's chart of 1612. It is in Commelijn's *Begin ende Voortgang vande Neederlandtsche Oost Indische Compagnie* (Amsterdam, 1646, 4to).

J. A. Colom's chart in *Der Vyerighe Colom* (Amsterdam, 1648).

Blaeu's chart of 1663 (?), published in *Le Grand Atlas ou Cosmographi Blaviane* (Amsterdam, 1663).

Pieter Goos' chart of 1666 (?), entitled *De Custen van Noorwegen, Finmarken, Laplandt, Spitsbergen*, &c., published at Amsterdam about 1666.

J. Van Keulen's chart, published in his atlas of 1689 (Amsterdam).

Giles and Reps chart, published at Amsterdam by Gerard Van Keulen, about 1707.

Zorgdrager's chart in his *Groenlandische visscherij* (Amsterdam, 1720, 4to).

W. Scoresby's chart in his "Account of the Arctic Regions" (Edinburgh, 1820, 8vo).

INDEX

[The numerous names in the Appendix, there arranged in geographical order, are not included in this Index.]

THE END

Printed by BALLANTYNE, HANSON & Co.
Edinburgh & London

Spotted something that isn't quite right?

Let us know and we'll fix it for future reprints. And as a thank you for your diligence, we'll send you a free Sothis book of your choice.

publisher@sothis.press

This page intentionally left blank

Except of course for that line above. And this one. And whilst we have your attention, perhaps you'd like to follow us on Instagram @sothis press. Go on.

Made in United States
Orlando, FL
05 November 2026

90229827R00298